LIFE SCIENCES AND SPACE RESEARCH X

COSPAR

LIFE SCIENCES AND SPACE RESEARCH X

Proceedings of the Open Meeting of Working Group 5 of the Fourteenth Plenary Meeting of Cospar

Seattle, Washington, USA, 21 June — 2 July, 1971

Organized by

THE COMMITTEE ON SPACE RESEARCH — COSPAR

and

THE UNITED STATES NATIONAL ACADEMY OF SCIENCES

Edited by

W. VISHNIAC

AKADEMIE-VERLAG · BERLIN

1972

Executive Editor:
Dr. A. C. Stickland

Library of Congress Catalog Card Number 63—6132

202 · 100/467/72
Gesamtherstellung: VEB Druckerei „Thomas Müntzer", 582 Bad Langensalza
Bestellnummer: 761 673 4 (3060/X) · ES 18 D 1, 18 E 2
Printed in German Democratic Republic

Preface

With the appearance of the Xth volume of this series, it seems instructive to see whether trends can be observed in the material which investigators have presented to their colleagues at COSPAR meetings during the past decade. The first volume appeared at a time when actual experience in space flight, manned or automated, was minimal. The emphasis was necessarily placed on interpretation, discussion, planning and speculation.

Since then a great variety of experience has been gained in earth orbit and on the moon, and it has become possible to define more closely the biological, and specifically human, problems which are encountered in space. With this experience it has also been possible to place the results in an order which allows us to develop a strategy for the systematic exploration of problems we encounter. The lines of research group themselves naturally as follows.

I. Physiology of living organisms in space; thus gravitational biology and radiation biology emerge as recurrent and well defined topics.

II. Instrumentation, the development of spaceworthy apparatus for automated or man-tended experiments, has become a highly specialized subject.

III. The protection of investigations from terrestrial influences ranges from the elimination of gravitational effects and earth-bound periodic events to planetary quarantine; the establishment of a barrier which would prevent terrestrial micro-organisms from interfering with projected planetary investigations. One corollary of the latter topic has been the quarantine precautions with which lunar samples have been surrounded, precautions which by COSPAR resolution are to be extended to all extraterrestrial matter returned to earth.

IV. There remains always the look towards the future, the planning of explorations which reach further into space.

Exobiology, sometimes called "a science without a subject", has thus come into being. Nevertheless, in a wider sense exobiology has become a laboratory subject and the final section of this volume is devoted to it.

This volume appears as three exploratory vehicles, Mars 2 and 3 and Mariner 9, have reached Mars, and another, Pioneer F, speeds towards Jupiter. Furthermore, during the past few years radioastronomers have found a hitherto unsuspected array of organic compounds in interstellar space. The biological significance of these studies is therefore rapidly expanding the scope of the life sciences into the further reaches of the solar system and beyond.

The editor wishes to take this opportunity to thank Dr. Stickland for her invaluable role in preparing the current volume for the press.

WOLF VISHNIAC.

Throughout this work, references to the various volumes of *Life Sciences and Space Research* are given in the form: volume number, page, date. The publishers of the volumes are as follows:

Vols. I–III, V–VIII,	North-Holland Publishing Co., Amsterdam;
Vol. IV,	Spartan Books Co., Washington, D. C.;
Vols. IX, X,	Akademie-Verlag, Berlin.

CONTENTS LIST

Joint Open Meeting of the Panel on Planetary Quarantine and Working Group 5

Life Sciences and Space Research X — Akademie-Verlag, Berlin 1972

DEVELOPMENT OF PLANETARY QUARANTINE IN THE UNITED STATES

D. G. Fox[a], L. B. Hall[a] and E. J. Bacon[b]

[a] NASA Planetary Quarantine Office, Washington, D. C., USA
[b] Exotech Systems, Inc., Washington, D. C., USA

This paper traces the development of the United States Planetary Quarantine Program, with emphasis on progress during the past four years. The National Aeronautics and Space Administration Planetary Quarantine Program closely follows policies recommended by the ICSU, COSPAR and the United States National Academy of Sciences Space Science Board. Policy formulation, program planning, and implementation follow an orderly process recommended by a Planetary Quarantine Advisory Panel and other related groups. In fulfilling its obligations, the Planetary Quarantine Office guides and supports quarantine activities within United States planetary flight projects and certifies projects for launch. In particular this office issues directives and standards and conducts research germane to the fulfillment of these constraints. The paper uses ongoing flight programs to illustrate the role of the NASA Planetary Quarantine Program in assuring the biological integrity of the planets.

1. Background

With the advent of space exploration, the possibility of infecting planets of our solar system with terrestrial organisms or of depositing materials which could confound scientific experiments has become a reality. This possibility was foreseen as early as 1957, when the National Academy of Sciences of the United States (NAS) expressed deep concern that contamination by early space explorations would endanger certain scientific investigations of the planets. NAS urged the International Council of Scientific Unions (ICSU) to assist in the evaluation of contamination hazards and encourage development of preventive measures. Subsequently, the ICSU formed an *ad hoc* committee on Contamination by Extraterrestrial Exploration (CETEX) [2] which in 1958 recommended the adoption of a code of conduct aimed at achieving a compromise between an all-out program of lunar and planetary exploration and a desire to provide maximum protection against degradation of future scientific studies.

CETEX in 1959 [3] took the position that the contamination problem was an integral part of the duties of the ICSU Committee on Space Research which had been established the preceding October. COSPAR assumed responsibility for consideration of the contamination problem and asked the United States of America and the Union of Soviet Socialist Republics to consider methods for avoiding the transfer of terrestrial organisms to the planets. In response to this concern the United States established a program of planetary quarantine and assigned it to the National Aeronautics and Space Administration for implementation. The objective of this program is to prevent the transfer of terrestrial life to the planets of biological interest so that life detection experiments will not be invalidated and the planet will not be overgrown by terrestrial life with consequent irreversible changes in its environment.

2. NASA Policy

NASA's planetary quarantine program is guided by a policy for the development and use of planetary quarantine standards. Based upon Article IX of the Space Treaty [4], this policy is continuously reviewed and modified as indicated by the work of pertinent groups and panels within the Council of Scientific Unions, the COSPAR, and by the recommendations of the Space Science Board of the NAS. The American Institute of Biological Sciences has established for NASA a Planetary Quarantine Advisory Panel, composed of microbiologists, mathematicians and engineers, to advise the Planetary Quarantine Office in program planning technology and evaluation. Policy provisions are promulgated to flight program organizations and NASA installations in a set of policy directives and provision documents. The three major quarantine policy directives are: 1. NPD 8020.7 Outbound Spacecraft: Basic Policy relating to Lunar and Planetary Contamination Control. 2 NPD 8020.10 Outbound Planetary Biological Contamination. 3. NHB 8020.12 Planetary Quarantine Provisions for Unmanned Planetary Missions. The first of these defines the responsibilities associated with the certification of the biological burdens on outbound automated spacecraft. The second deals with the control of these bioburdens to comply with contamination control standards. The third sets forth planning and control requirements applicable to flight programs in the fulfillment of their quarantine obligations.

These directives are issued so that all NASA automated planetary flight programs will be governed by a uniform set of planetary quarantine requirements. Satisfaction of these requirements by flight projects will enable NASA management to assure organizations concerned that this nation has established biological safeguards in its planetary programs and that a course is being followed which will assure that the planets remain as biological preserves for scientific investigations. Under this system all flight projects must demonstrate compliance with specified quarantine requirements before launch approval is granted by the Planetary Quarantine Officer. Each project is required to perform an analysis to show that the precautions taken in the design and implementation of the flight project result in a probability of contamination less than a value specified by the Planetary Quarantine Office in keeping with the overall allocations recommended by COSPAR.

3. Planetary Quarantine Implementation

NASA's procedure for the establishment of quarantine constraints and the analysis which demonstrates compliance employs three elements: (i) mathematical models of the principal parameters and their interrelation; (ii) an administrative division of responsibility for the establishment of parameters between NASA Headquarters and the flight project, and (iii) the selection and justification of quantitative values for each parameter.

Mathematical models have been developed for planetary quarantine which insure the identification and orderly interrelation of each potentially contaminating factor. The resulting models are used to apply quarantine constraints and to evaluate flight project quarantine plans. The model used to determine the risk constraint $P(N)$ for each mission uses the COSPAR established limit P_C for the probability of contaminating the given planet, the estimated number of missions N to be conducted to the planet during the period of biological interest and an allocation policy based upon the different types of missions such as landers, orbiters, lander–orbiter combinations and flybys:

$$P_C = N_1 P(N_1) + N_2 P(N_2) + \cdots$$

P_C = Probability of contaminating given planet
N = Number of missions of each type assumed to be conducted
$P(N)$ = Probability that mission contaminates the planet (mission risk constraint).

Prior to 1966, $P(N)$ values for planetary missions were established by COSPAR specifying two general constraints, namely the probability that a lander contains a viable micro-organism, and the probability of accidental impact by an unsterilized non-lander. By 1966, however, COSPAR had decided that the detail of mission allocation is best left to the nation responsible for the mission. Accordingly, allocations for United States flight programs are now established by the NASA Planetary Quarantine Office, in accordance with general guidelines established by COSPAR and the Space Science Board. $P(N)$ values for ongoing United States missions are given in Table 1. The difference in values for the

Table 1
NASA mission constraints for ongoing US planetary flight programs

Mission	Planet	$P(N)$
Mariner 1971	Mars	7.1×10^{-5}
Viking 1975	Mars	7.2×10^{-5}
Pioneer F/G	Jupiter	6.4×10^{-5}
Mariner 1973	Mercury	7×10^{-5}
	Venus	7×10^{-5}

two Mars missions of 1×10^{-6} is due to the lander capsule for Viking. Values for Jupiter and Mercury reflect different total numbers of missions estimated for these planets.

A second model is employed for administrative sub-allocation of the mission risk constraint $P(N)$ among the major mission components such as the launch vehicle, the bus, the lander, and components that may be jettisoned during trans-Mars cruise. For each sub-allocation the applicable potential contamination sources are identified and assessed for contamination probability. Some of these sources and their related major mission components are shown in Table 2.

Table 2
Sources of contamination

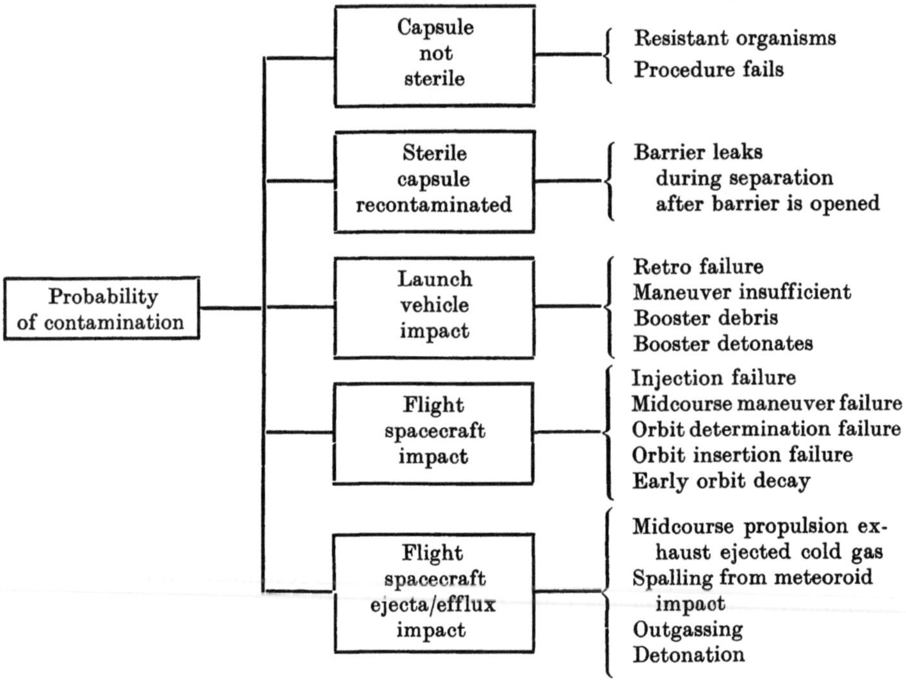

Landing hardware could arrive on the planet in a contaminated state due to sterilization inadequacies, leaks, or recontamination from the bus or orbiting spacecraft. Accidental impact of vehicle parts not intended to land and hence not sterile, could also result in contamination of the planet. A possible source which deserves careful scrutiny is the dislodgement or release of micro-organisms from the spacecraft and subsequent arrival in the planetary atmosphere or on the planet's surface. Mechanisms for such events include micrometeoroid dislodgement, expulsion with attitude control gases and with the firing of pyrotechnic devices. An accounting of all potential sources of contamination is essential for accurate estimation of the probability of contamination from space flight hardware that has not been sterilized and contains viable organisms. This is accomplished with the aid of a third set of models.

In the case of orbiting or flyby missions, the analysis includes examination of the hazards associated not only with accidental impact of the spacecraft, but with all other events which may release organisms from the spacecraft. The model systematically accounts for such factors as the probability of the occurrence of each source, the probability that viable organisms released reach the planet, that the micro-organisms survive the trip, and are released on the planet and the probability that the micro-organisms grow.

A special case model applicable to landing missions relates the mission contamination risk to the number of buried micro-organisms $m_B(O)$, permitted in the spacecraft. The model accounts for the destruction of the bioburden due to lethal stresses. It can be used to determine sterilization requirements when values are assigned for the probability of release, $P(r)$, and the probability of growth and proliferation, P_G. The measure of microbial thermal resistance is the decimal reduction time D, defined as the time required at a given temperature to destroy 90% of a microbial population. Time, t, is the duration of dry heat sterilization imposed upon the spacecraft material.

$$P_C(B) \geq m_B(O)\ 10^{-t/D_B}\ P_B(r)\ P_G$$

$P_C(B)$ = allocation to buried bioburden
$m_B O)$ = initial number of buried micro-organisms
t = sterilization time
D_C = decimal reduction time for buried burden
P_G = probability of growth

Similar models treat the surface and mated biological burdens and can be used to establish the needs for contamination control and sterilization during assembly of spacecraft components.

The utilization of mathematical modeling has resulted in mission constraints which, despite uncertainties in some parameter values, can be satisfied by flight projects and be substantiated to be responsive to the terms of international agreements.

4. Program Office Contamination Constraints

A difficult task in evaluating the contamination probability of a proposed mission is the selection of quantitative values for some of the model parameters. This process may involve a considerable amount of subjective judgment by the most knowledgable experts in the field in combination with results of research and data from earlier space exploration. This process entails necessary risks which will be reduced as information is gradually accumulated and evaluated by authoritative persons and groups in the field. Although not fully satisfactory, at the present time it is the only method available for assigning values to some parameters and has the desirable characteristics of focusing attention on each individual item rather than on broad groups of items.

The number of micro-organisms encapsulated in materials or residing between mated surfaces cannot currently be determined with precision owing to the inefficiency of sampling processes. The uncertainty may be one or more orders of magnitude. Estimates run from less than 1 per cm^3 to more than 1000 per cm^3 [5], though with the heating that is applied to most of the piece parts to be used the true value is almost certainly never the upper limit. With these densities the microbial load buried in a Viking type lander could range from less than 1×10^5 to 1×10^9. This implies an uncertainty of more than 20 hours in sterilization time. We are currently sponsoring research [6] to reduce the uncertainty in the value of the buried load and to establish a number which the flight project can confidently use without risk of excessive safety margins.

Our current lack of knowledge of planetary conditions is reflected in the wide range assigned to the probability that the planetary environment can support growth and proliferation of the terrestrial organism (see Table 3). Values

Table 3

Value for probability of growth P_G

Planet	P_G
Mercury	1×10^{-6}
Venus (surface)	1×10^{-6}
Venus (atmosphere)	1×10^{-4}
Jupiter, Mars and all other planets	1×10^{-4}

for the probability of growth have been considered and recommended by the Space Science Board. These values are under constant review by the Planetary Quarantine Office to ensure that they reflect latest knowledge of planetary conditions and are not needlessly conservative. A recent reappraisal of the latest data on Mars by the Space Science Board [7] led to a reduction of P_G for Mars from 10^{-3} to 10^{-4}. Re-evaluations are also expected for Venus, Mercury and Jupiter during the coming year.

D-values (Table 4) are based upon laboratory sterilization research on dry heat resistance of micro-organisms. Early in the United States program,

Table 4

D-values currently recommended by the Planetary Quarantine Advisory Panel

D_{125} (encapsulated)	5 hours
D_{125} (mated)	1 hour
D_{125} (surface)	0.5 hour

D-value is the time required at a stated temperature to destroy 90% of a biological population.

Bacillus subtilis var. *niger* was designated as an appropriate test organism for dry heat sterilization. It is well known, however, that microbial resistance to dry heat varies greatly according to the environmental characteristics of the materials in and on which the organisms are located during sterilization [8]. A spacecraft contains a variety of materials with differing thermal and moisture diffusion characteristics. Even when a judicious classification is made, significant efforts are needed to categorize microbial resistance. At present, we are aware of at least an order of magnitude range in resistance to dry heat sterilization for the same test species depending upon whether they are on the open surface of the material, located between two mated surfaces, or encapsulated within a material.

Attention is currently focused on an important environmental parameter, namely, the water activity in and around the spores during sterilization [9]. That this factor greatly influences microbial dry heat resistance is already well established. Water activity may, in theory, explain the variations in heat resistance, but it remains to be determined in what manner it should enter the specifications for the spacecraft sterilization process. The Planetary Quarantine Office has the responsibility to determine and approve values for these factors. It employs groups (the Space Science Board and the Planetary Quarantine Advisory Panel) to achieve this end. These specialists also assist NASA to formulate research by university, industry and government laboratories which may result in possible relaxation in these constraints. Flight projects are free (and encouraged) to review values specified by the Planetary Quarantine Office and can submit for assessment and approval other values which their analyses can substantiate. In general, such deviations apply to mission specific factors where the impact of alternative sets of parameter values on project operations are more directly felt.

The Planetary Quarantine Office also specifies to flight projects the value of $P(\mathrm{r})$, the probability of releasing a viable organism from the spacecraft. In the past we have used a conservative value of unity for the probability of release of buried micro-organisms but according to recent results of sponsored research a smaller value can be substantiated. We have investigated two release mechanisms [10] pertinent to Martian missions. These are: (i) release due to material fracturing in high velocity impact of the space hardware, and (ii) release from hardware landed on the planet's surface as a result of aeolian erosion over an extended period of time. Impact releases less than 1% of encapsulated burden in a viable state regardless of the amount of fracturing; high impact velocities which create extensive material fracturing also cause significant lethality [11]. Laboratory tests [11], have demonstrated that aeolian erosion is a significant release mechanism. We are currently attempting to estimate the degree of erosion to be experienced during the period of biological interest by a Martian lander in an attempt to establish a highly realistic value of $P(\mathrm{r})$.

The selection of values for some of the parameters involves a considerable amount of judgment. The proportion of factual information to expert speculation has increased as research results and flight data yield more useful information. The need for estimation is expected to continue for some years to come; when experimental evidence does not exist we will continue to obtain a concensus of the most authoritative minds in the field of knowledge that is applicable to supplement experimental evidence. Parameter determination is an iterative process. To assist in this task the Planetary Quarantine Office is preparing documentation [12] intended to assist flight projects in the required analysis. An illustration of the contamination analysis of a non-landing mission is that of the recently launched Mars Mariner 1971 orbiter mission [13].

5. Sterilization Technology

Progress in spacecraft sterilization technology has been characterized by an ever increasing pace from statements of objectives to the design of methodology for implementing those objectives. The sterilization technology available when the spacecraft sterilization effort was started was essentially that which could be borrowed from the pharmaceutical, surgical supply and canning industries. It has been necessary to extend the technology greatly to apply it to complex spacecraft projects. The development of implementation procedures appropriate to established goals involves arduous tasks for microbiologists and engineers alike; significant progress has been achieved, but much is yet to be done if sterilization stresses are to be fully effective, but not excessive. Although, dry heat is currently the only approved spacecraft sterilization method, we are continuing research to develop other sterilization and decontamination techniques, including ethylene oxide [14], paraformaldehyde [15], and thermoradiation [16]. The Viking '75 program which will soft land an exploratory vehicle on the surface of Mars in mid 1976, represents a milestone in our Planetary Quarantine program. The Viking spacecraft will be the first United States planetary vehicle subjected to rigorous dry heat sterilization. The plan for fulfilling the quarantine requirements on this program has been prepared by the Viking project team and we have every assurance that these requirements will be satisfactorily met without adverse effects on spacecraft reliability or incurring excessive program costs.

Other current supporting research and development is directed toward improving our understanding of quarantine needs and our knowledge of applicable factors and parameters so that requirements for future flight projects will not impose unrealistic burdens for contamination control and sterilization. As our knowledge increases, we are able to effect relaxations in mission constraints enhancing our confidence that compliance with future constraints will becomes less difficult and the United States Planetary Quarantine Program will fully meet the international obligation to protect the planets of the solar system.

References

[1] L. B. Hall, in: COSPAR Technique Manual No. 4, 1968 (p. 3).
[2] Committee on Contamination by Extraterrestrial Exploration (CETEX), Science **128**, 887 (1958).
[3] CETEX, Nature, Lond. **183**, 925 (1959).
[4] United States Treaties and Other International Agreements, Vol. **18**, 2410 (1967).
[5] C. W. Craven, et al., NASA SP 108, 1966 (p. 47).
[6] R. G. Lyle, Exotech Systems, Inc, personal communication, June 1971.
[7] National Academy of Sciences Space Science Board, Space Science Board Meeting, Woods Hole, Mass., July 1970.
[8] D. Vesley et al., Univ. Minnesota Semi-Annual Progr. Rep. No. 4, June 1970 (p. 57).
[9] R. Angelotti, in: COSPAR Technique Manual No. 4, 1968 (p. 59).
[10] S. Schalkowsky, Exotech Systems, Inc. Tech. Rep. 71-11, Dec. 1970.
[11] R. Green, Jet Propulsion Lab., personal communication, June 1971.
[12] Exotech Systems, Inc., Task 7, Contract NASw-2062, work in progress.
[13] R. Green and A. R. Hoffman, Jet Propulsion Lab., personal communication, June 1971.

[14] R. G. LYLE, Exotech Systems, Inc., personal communication.
[15] F. BEYERLE, Becton, Dickinson Research Center Tech. Rep. on contract NAS8-24513, Dec. 1970.
[16] H. D. SIVINSKI et al., Sandia Labs. Tech. Rep. on contract NASAw-12,853, Dec. 1970.

Life Sciences and Space Research X — Akademie-Verlag, Berlin 1972

MICROBIOLOGY OF THE DRY VALLEYS OF ANTARCTICA

N. H. Horowitz[a] and R. E. Cameron[b]

[a] California Institute of Technology, Pasadena, Calif., USA
[b] Jet Propulsion Laboratory, Pasadena, Calif., USA

The ice-free valleys of South Victoria Land, Antarctica, are the coldest and driest deserts on the earth. As such, they provide a unique opportunity for biological studies in an extreme and in some ways Mars-like natural environment. The life of the valleys is almost entirely microbial. Dense populations of algae, fungi, bacteria, and protozoa are found where favorable conditions exist, for example, on the shore of saline lakes formed by glacial melt in the lowest part of some of the valleys. In the drier parts of the valleys the microbial populations are very small, however, and some soil samples are sterile. The observations suggest that the dry valleys are an essentially abiotic region in which the small microbial population is maintained by fallout of cells blown from favorable locales such as the lake shores and the ocean. Little cell multiplication can be occurring in the driest parts of the valleys. One implication is clear: the possibility that terrestrial micro-organisms can grow in the far more hostile environment of Mars is negligible.

The paper will be published in full.

Life Sciences and Space Research X — Akademie-Verlag, Berlin 1972

PLANETARY QUARANTINE ANALYSIS FOR AN UNMANNED MARS ORBITER

A. R. HOFFMAN[a], R. J. REICHERT[a], N. R. HAYNES[a] and L. B. HALL[b]

[a] Jet Propulsion Laboratory, Pasadena, Calif., USA
[b] NASA Headquarters, Washington, D. C., USA

The United States Mariner Mars 1971 program was planned to place two planetary vehicles in Martian orbit to obtain scientific measurements of the physical characteristics of the planets. To ensure that Mars will be maintained as a biological preserve, the probability of contaminating the planet with viable terrestrial micro-organisms from the launch vehicle or the spacecraft has been analyzed.

This paper describes the analysis approach and the planetary quarantine model being applied to the program for allocating and estimating the probability of contamination associated with potential biological contamination sources. It is shown that three sources — accidental impact of the spacecraft, loose particles, and gases used for attitude control and pressurization — form the major hazards. Furthermore, the results of the analysis indicate that with the planned mission strategy, including aiming point and delivery biases, and the imposition of facility and procedural control during the systems test operations to minimize particulate and microbial contamination of the spacecraft, the planetary quarantine contraints for the Mariner Mars 1971 mission are being satisfied.

1. Introduction

The United States' concern for the biological preservation of the planets, in consonance with COSPAR agreements and the Outer Space Treaty [1], necessitated the development of planetary quarantine requirements, policies, and approaches for planetary flight programs [2]. This digest will identify the considerations and approaches that are used for the Mariner Mars 1971 orbital mission to assure concerned organizations that the United States is pursuing an active planetary quarantine program.

The United States policy is established by the National Aeronautics and Space Administration (NASA), following the recommendations of the Space Science Board of the National Academy of Sciences. The objective of the NASA policies is to prevent the transfer of terrestrial life to the planets of biological interest so that life detection experiments will not be invalidated and the planet's environment will not be irreversibly altered. The Planetary Quarantine Office of NASA implements national policy for unmanned planetary missions by means of a planetary quarantine document [3] that requires all flight projects to demonstrate compliance with specified quarantine constraints before launch approval is granted. Each planetary project performs an analysis to show that the probability of contamination associated with a flight will be less than a specified value. The value specified by the Planetary Quarantine Office for each mission is established from the planet contamination constraint accepted from COSPAR, which is based on the estimated number of missions to that planet to be conducted during the period of biological interest, the proportion of United States launches, and the apportionment of the total constraint against the various types of mission (i.e. lander, orbiter, flyby). The mission allocation for Mariner Mars 1971 is 7.1×10^{-5}. In performing the planetary quarantine analysis, the project utilizes NASA directives, provisions, and standards. Among the latter is a value for probability of growth and proliferation for any micro-organisms reaching Mars, $P_G = 1 \times 10^{-4}$.

2. Mission Profile

The primary objective of the Mariner Mars 1971 Project is to insert spacecraft into orbit about Mars to obtain scientific information about the planet and its environment, and to make exploratory investigations that will form the basis for future experiments, specifically those relevant to the search for extraterrestrial life. The Mariner Mars 1971 spacecraft was launched by an Atlas Centaur launch vehicle from Cape Kennedy. Flight time to Mars is 168 days. The mission profile for the Mariner Mars 1971 Program is given in Fig. 1. The mission profile can be defined in two segments, the heliocentric

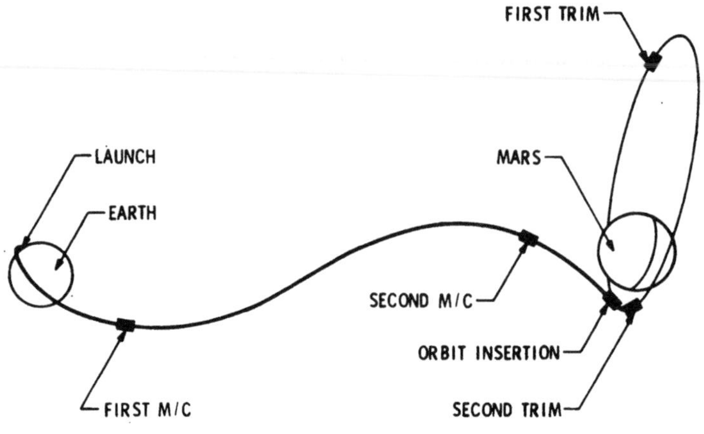

Fig. 1. Mission profile. (M/C = mid-course.)

and areocentric. The heliocentric portion of the mission is from injection to, but not including, orbit insertion. The areocentric portion takes place after and including orbit insertion.

3. Planetary Quarantine Model and Analysis

From the mission profile and from a flight sequence of events, all potential contaminating sources are identified. A planetary quarantine model (Fig. 2) is then determined. This probability "tree" permits estimation of constraint levels as it relates to each identified source. Confirmation that these estimates will not be exceeded is obtained by performing an analysis utilizing information from prior flight experience, literature surveys, experimentation, and engineering judgment. The results of this analysis are used for selecting appropriate mission strategies, i.e. aiming point biasing and orbit periapsis altitude selection.

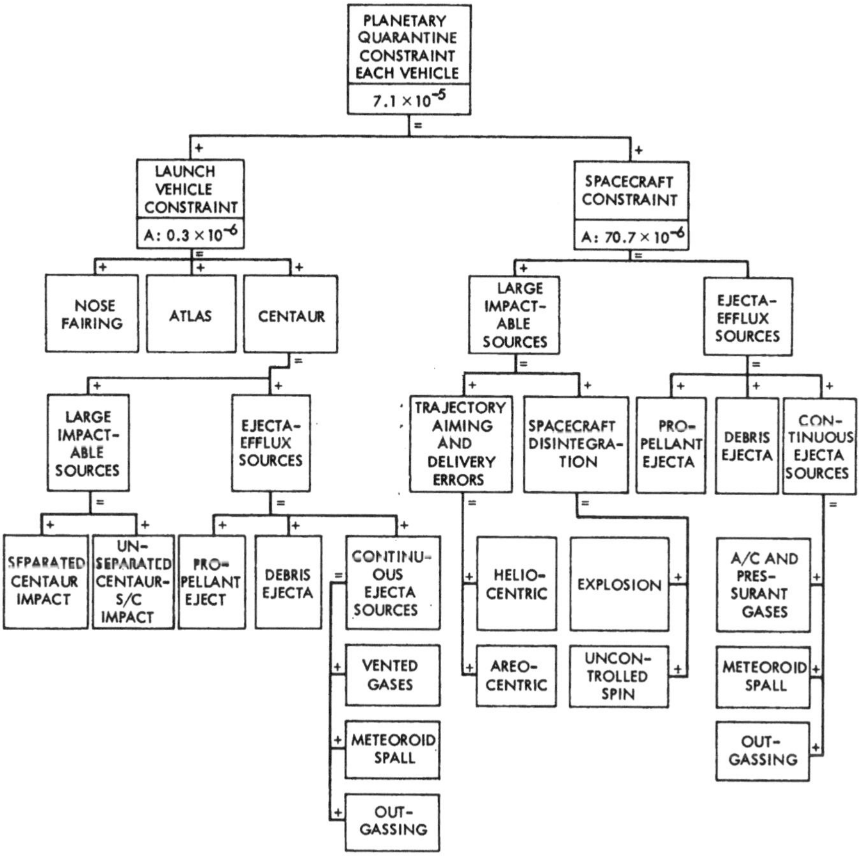

Fig. 2. Planetary quarantine model.

Two categories of contamination sources are identified: large impactable sources and ejecta efflux sources. Examples of the first category are the launch vehicle, spacecraft, medium-gain antenna plug or large pieces that could impact the planet as a result of trajectory aiming errors, orbital decay, and catastrophic events during the quarantine period. The second category includes fluids or particles that consist of a relatively small amount of material expelled at any time.

The spacecraft portion of the model is considered in more detail in Fig. 3. The numbers under each source are the prelaunch allocations. The "trajectory aiming and delivery errors" source considers the accidental impact of the

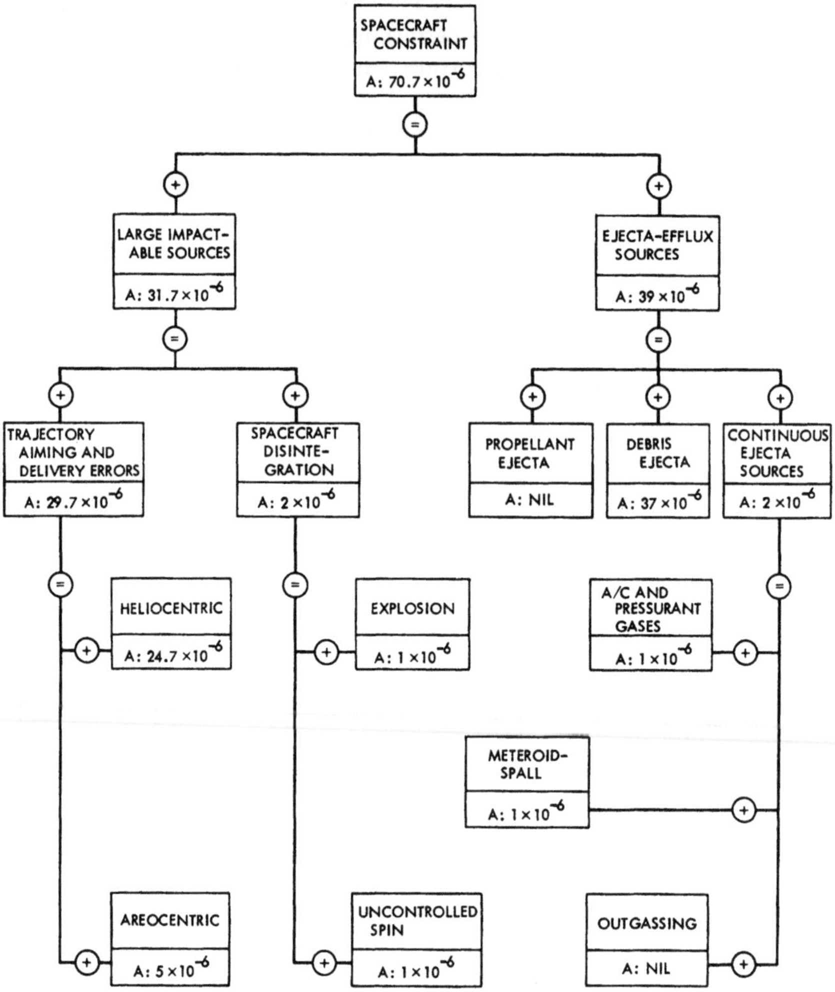

Fig. 3. Planetary quarantine model — spacecraft.

spacecraft as a result of injection, orbit determination, and maneuver execution errors and subsequent delivery dispersions during the heliocentric and areocentric phases of the mission. These probability allocations are met by mission and navigation strategy, including aiming point and delivery biases. Catastrophic events that could occur while the spacecraft is in orbit include spacecraft disintegration by explosion and uncontrolled spin. The probability of contamination by spacecraft disintegration was analytically determined by calculating the probability that an explosion or uncontrolled spin would occur and that the event would result in placing a large equipment item into a decaying orbit or impact trajectory.

There are three major ejecta efflux sources: propellant, debris, and continuous. Debris ejecta, including loose particles, is the principal contamination source; thus the largest suballocation is given to it. All of these sources were analyzed in detail. The probability values that were used in the calculations are summarized in Table 1. The estimate for the number of organisms in or

Table 1
Probability values

Parameter	Estimate	Information source
Number of organisms	Source dependent	Encapsulation bioassay expected die-off
Probability of ejection	Source dependent	—
Probability of surviving ejection*	1.0	Engineering judgment
Probability of impact	Source dependent	—
Probability of surviving space environment	1×10^{-3}	UV experiments in literature
Probability of surviving atmospheric entry	1×10^{-2}	Engineering judgment
Probability of release	1.0	—
Probability of growth	1×10^{-4}	NASA directive

* With meteoroid bombardment, 5×10^{-1}.

on a given ejecta efflux source is dependent on the source that is being analyzed. For example, the item of interest for the debris ejecta is the number of organisms on the exposed surface of the spacecraft, whereas for the gas efflux an estimate of the number of organisms in the on-board nitrogen gas bottles is appropriate. The probability of ejection P_E of a single micro-organism and the probability P_T that the organism is placed on an impact trajectory are also source-dependent. An example of the manner in which these source-dependent parameters were estimated is shown in Fig. 4 for meteoroid spall. The estimate corresponds to the largest value of $P_E \times P_T$ over the expected range of ejecta size and density. The probability that an organism will survive

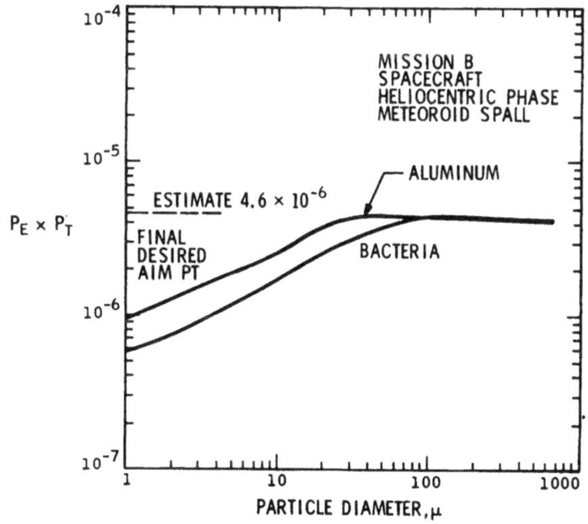

Fig. 4. Example of source dependent parameter, $P_E \times P_T$.

ejection is assumed to be 1.0 for those ejection processes involving cold release. Organisms released through meteoroid impact are subjected to some heat. Consequently, using engineering judgment, the estimate for meteoroids is 0.5.

The probability of survival of a micro-organism in the space environment was a difficult parameter to estimate because none of the studies in the literature could be directly related to the Mariner Mars 1971 mission conditions. The probability of survival value is assumed to be a function of the amount of time the particle is exposed to the environment after release from the spacecraft. The environmental exposure, whether during the heliocentric or areocentric portions of the mission, will be for a minimum of several days and may extend for months. Based on the results of ultraviolet experiments in published literature, the time of expected exposure, conferences with experts in the field, and engineering judgment relative to the protection afforded by material surrounding an organism, the probability of survival was estimated to be 1×10^{-3}.

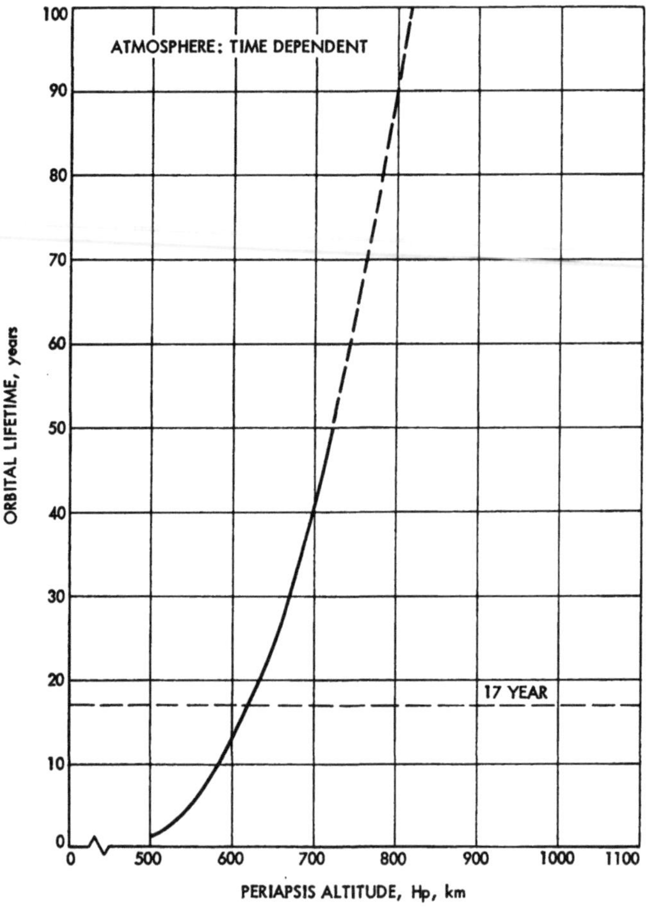

Fig. 5. Spacecraft orbital lifetime. Orbital period, 12 h; inclination, 65°.

The probability that a micro-organism will survive atmospheric entry depends upon the size and emissivity of the particle on which it resides. Particles that encounter the atmosphere will do so at relatively high velocities. Depending upon particle parameters, temperatures sufficient to produce sterilization could be achieved. An estimate of 1×10^{-2}, based on engineering judgment, was selected. The probability of release was assumed to be 1.0. The probability of growth was given a value of 1×10^{-4} by directive from the NASA Planetary Quarantine Office in accordance with the recommendations of the Space Science Board.

The principal hazards from a planetary contamination standpoint are spacecraft impact, debris ejecta, and attitude control and pressurant gas efflux. During the heliocentric portion of the mission, maneuvers are performed in such a fashion that the probability estimates are within the probability allocations for spacecraft accidental impact events. For the areocentric portion of the mission, the orbit is chosen with a periapsis altitude such that the spacecraft will remain in orbit (Fig. 5), even after mission completion, for much more than the duration of the period of biological exploration, which for Mars is defined as being through 1 January 1989, 17 years after insertion. Debris ejecta, including loose particles released by spacecraft dynamic events or meteoroid impact, are a hazard near encounter and during the areocentric

Fig. 6. Spacecraft laminar downflow tent.

portion of the mission. To meet the planetary quarantine constraint for this source, the spacecraft was carefully cleaned prior to launch. The attitude control and pressurant gases are a potential principal hazard. However, during the filling process for Mariner Mars 1971, the gases are filtered through 0.5 μm filters into precleaned tanks.

4. Microbial Burden Control and Estimation

Based on the planetary quarantine analysis, an upper permissible microbial burden level was established for the spacecraft at the time it was placed within the nose fairing, i.e. encapsulation. For a nominal mission, this level was 1×10^5. To ensure that the permissible level would not be exceeded, the spacecraft were assembled, tested, and encapsulated in Class 100 laminar downflow tents (Fig. 6), clothing and access restrictions for personnel were established, and an extensive cleaning program using isopropyl alcohol on critical spacecraft surfaces was implemented. Microbiological assays (Fig. 7) were taken using the swab rinse method [4]. The United States Public Health Service verified the assays. The estimated microbial burden on the exposed surfaces of the Mariner 9 spacecraft was 3.1×10^4. This estimate is the lowest that has been measured on any Mariner spacecraft, indicating the effectiveness of the cleaning and handling program.

Fig. 7. Microbiological sampling of the Mariner 1971 spacecraft.

5. Summary

Satisfaction of the planetary quarantine requirements by all United States planetary flight programs enables NASA to assure concerned organizations that this nation has established biological safeguards in its planetary programs and that a course is being followed that will result in the planets being manitained as biological preserves for scientific investigations.

The United States unmanned Mars orbiter project has analyzed the probability of contaminating Mars with viable terrestrial micro-organisms carried on or ejected from the launch vehicle or spacecraft. A mathematical model has been constructed to allocate and to estimate probability of contamination associated with identified contaminating sources or events. Mission strategy, including aiming point biasing and orbit periapsis altitude selection, was developed to satisfy the probability allocations for accidental spacecraft impact. To ensure that permissible microbial burden levels would not be exceeded, extensive cleaning and facility personnel control programs were implemented. The analysis and microbiological assay results indicate that the planetary quarantine constraints for the orbiter mission are being satisfied.

Acknowledgments

The support of M. Christensen and R. Koukol of the Jet Propulsion Laboratory, P. LaLime and G. Simko of AVCO, and N. Fields of the United States Public Health Service during the Mariner 1971 Planetary Quarantine Program, specifically for their cleaning, contamination control, or microbiological monitoring efforts, is deeply appreciated. Also, the cooperation and support of the spacecraft test and operations team is gratefully acknowledged, especially V. Ohanesian, T. Laney, T. Shaw, R. Forney and J. McGee.

This paper presents the results of one phase of research carried out at the Jet Propulsion Laboratory, California Institute of Technology, under Contract No. NAS 7-100, sponsored by the National Aeronautics and Space Administration.

References

[1] Treaty on Principles Governing the Activities of States in the Exploration and Use of Outer Space, TIAS 6374, 27 January 1967.

[2] D. G. Fox, L. B. Hall and E. J. Bacon, Life Sciences and Space Research X, 1 (1972).

[3] NASA, NHB 8020.12, April 1969.

[4] NASA NHB 5340.1A, October 1968.

Life Sciences and Space Research X — Akademie-Verlag, Berlin 1972

A RE-EVALUATION OF MATERIAL EFFECTS ON MICROBIAL RELEASE FROM SOLIDS

D. M. TAYLOR[a], S. J. FRASER[b], E. A. GUSTAN[b], R. L. OLSON[b] and R. H. GREEN[a],

[a] Jet Propulsion Laboratory, Pasadena, Calif., USA
[b] The Boeing Company, Seattle, Wash., USA

A previous report concerned with the percentage of microbial release from the interior of solid materials after hard impact, raised questions about the possibility that dissimilar materials might have different release properties. Therefore additional studies were conducted to obtain information on the release of micro-organisms from different solid materials impacted onto two types of surfaces. The combined study was performed by inoculating 10^4 *Bacillus subtilis* var. *niger* spores into Eccobond and methyl methacrylate. These materials were then machined into projectiles and fired from guns into stainless steel plates or sand at velocities ranging from 168 to 1554 m s^{-1}. Bacteriological examination of the fractured particles was conducted to establish the number of viable spores released from the interior of the projectiles.

Analysis of the results from two solid materials, two impact surfaces and four velocities showed that the number of micro-organisms released is less than 1% in all cases. However, statistical evaluation of all data demonstrates a significant difference in percentage microbial release between materials. Since significant differences were observed between materials, broad extrapolations of percentage release data should be avoided until release characteristics of different classes of spacecraft solid materials have been determined.

1. Introduction

Quantitative determinations are required in any meaningful attempt to evaluate the danger of contaminating other planets by terrestrial probes. Estimates of the probability of release of micro-organisms from the interior of spacecraft solids by fracture due to impact is an essential factor to be evaluated in establishing planetary quarantine and spacecraft sterilization requirements. A previous report [1] concerned with microbial release from solids, presented empirical impact data that indicated that dissimilar solid materials might have different release properties. Therefore, additional studies were initiated to clarify and supplement those findings. This paper presents a refinement of the microbial release parameter in the light of new data recently obtained and discusses the effects of materials on the release of organisms from the interior of two different solids after fracturing on to two types of impacting surfaces. The significance of these data in terms of effect on a lander sterilization process is also discussed.

2. Technical Discussion

The impact studies were designed to provide comparative release data on two different solid materials impacted onto a hard and a soft surface. The solids selected were Eccobond, an adhesive amine-cured epoxy, and methyl methacrylate, a molding acrylic. The test materials were inoculated with controlled numbers of *Bacillus subtilis* var. *niger* spores. After polymerization, the seeded solids were machined into 1 g cylinders. All of the cylinders were weighed, surface sterilized, and stored in sterile tubes. The impact tests were conducted by using a gun to fire the seeded cylinders (Fig. 1). Four firing velocities were chosen that encompassed the ranges that would occur during Martian direct entry or from orbit. The lowest velocity in the test, 168 m s^{-1}, was obtained with compressed air. The other velocities of 457, 945, and 1554 m s^{-1} were obtained by using powder charges as the propellant. Impact

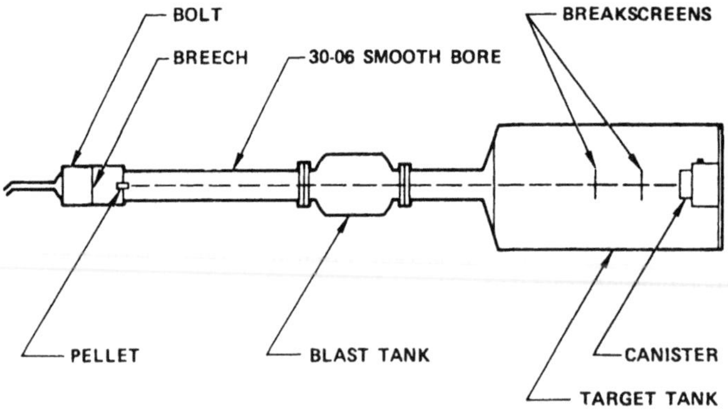

Fig. 1. Gas and powder gun.

of the seeded solids occurred on stainless steel or in sand. Microbiological analysis was performed on the recovered fragments by embedding the pieces in trypticase soy agar. The number of spores released from the interior of a solid upon material fracturing was calculated as a percentage of the number of spores initially available for release in an unfired solid. A detailed discussion of these laboratory procedures is given in [1]. The results reported there indicated that when Eccobond was impacted on to stainless steel, the percentage of spores released was approximately ten times greater than the number of spores from methyl methacrylate under the same conditions. This difference was attributed to a material effect or to differences in assay procedures, or a combination of the two factors. Since there is not a suitable solvent for Eccobond, grinding was the only method readily available for determining the initial number of organisms present in pellets before firing, while initial numbers in the methyl methacrylate were determined by dissolving the pellet. A gross visual examination of the fractured Eccobond pellets showed generally similar characteristics with respect to fracture patterns, surface areas and particle size after impact. The minor differences that were observed in fracturing did not appear great enough to account for the differences observed in percentage of spores released from the two materials. Therefore, the assay techniques were investigated, i.e. grinding of Eccobond and dissolving methyl methacrylate.

To compare the assay techniques, methyl methacrylate and Eccobond pellets were inoculated and fabricated so that each pellet contained approximately 5×10^4 *Bacillis subtilis* var. *niger* spores per gram of finished material. Recovery of the spores was obtained from individual methyl methacrylate pellets by utilizing both techniques, dissolving and grinding, to determine the effect of the abrasive grinding and encapsulation (those organisms that remain encapsulated in small particles after grinding). Particle size following the grinding was compared with Eccoband particles, and a relative efficiency factor was assigned to each technique. A relative factor of 1.5 for methyl methacrylate polymerization die-off factor and 20 for Eccobond was determined. This would mean that the initial spore count in pellets of Eccobond used in the impact studies determined by grinding should be multiplied by a factor of 20.

3. Results and Discussion

Recent data by Gustan and Olson [2] have been obtained for the percentage release of spores from inoculated Eccobond pellets impacted into sand at four velocities. These data, plus the data reported in [1] for methyl methacrylate impacted into stainless steel and sand and Eccobond into stainless steel at four different velocities are given in Table 1. It should be noted that all data reported for Eccobond have been adjusted by the relative grinding factor discussed above. An overall statistical analysis was performed on the data to determine the relationship between velocity, material, and impacting surfaces. The analysis indicates that velocity, materials, and impacting surfaces all have an effect on the percentage of spores released. In addition, the analysis indicates that there are highly significant interactions between velocities and impacting surfaces and velocities and materials, but no significant interaction between

Table 1

Percentage spores released at impact

	MM/SS				MM/S	E/SS	E/S
Initial spore level	10^2	10^3	10^4	10^5	10^4	10^4 [a]	10^4 [a]
Impact velocity ($m\,s^{-1}$)							
168	0.6 [b]	0.1	0.2	0.03	0.001	0.31	0.01
457	0.4	0.6	0.6	0.3	0.001	0.98	0.29
945	0.6	0.8	0.1	0.02	0.06	0.64	0.23
1554	0.2	—	—	0.002	0.06	0.06	0.34

MM/SS = methyl methacrylate onto stainless steel. MM/S = methyl methacrylate onto sand. E/SS = Eccobond onto stainless steel. E/S = Eccobond onto sand.

[a]) Initial inoculation levels adjusted relative to grinding factor ($\times 20$).

[b]) Mean of 6 replications.

impacting surfaces and materials. It shows that more spores are released at 168, 457 and 945 m s^{-1} from impacting stainless steel than from sand, but at the highest velocity (1554 m s^{-1}) fewer spores are released from the stainless steel impact (Table 2). The inverse relationship at the higher velocity may be due to the lethality associated with heat generated by the stainless steel impact. The highest release percentages for both materials occurred at 457 m s^{-1} when

Table 2

Effect of impact surfaces and velocity on percentage spore release

Impact velocity (m s^{-1})	168	457	945	1554[a]
Impact surface				
Stainless steel	0.24	0.80	0.36	0.03
Sand	0.008	0.15	0.14	0.19

[a] Data for methyl methacrylate at this velocity extrapolated to 10^4 initial inoculum level from 10^5 data.

impacted into stainless steel; the same effect was not observed for the sand impact tests. The interaction of velocities and materials indicate that the percentage of spores released from Eccobond is significantly higher except at the lower velocity (Table 3). In addition, the lethality due to impact occurs at a higher velocity for Eccobond than methyl methacrylate.

The overall objective of these studies was to determine the percentage microbial release from the interior of spacecraft materials. A review of all data from these studies indicates that, under the conditions simulated, less than 1% of internal spore contamination will be released due to impact. There are, however, other factors that may affect this number as it is related to a release probability for impact of a spacecraft into the surface of a planet. The first factor is the effect of initial inoculum levels. The effect of inoculum levels was studied only when methyl methacrylate was impacted into stainless steel, and the data (Table 1) indicate that as the inoculum level is reduced, the percentage of spores released is significantly increased. If this relationship exists for Eccobond, the data would indicate that at 945 m s^{-1} with lower inoculum levels, the percentage of spores released under these conditions will exceed 1%. The second factor for consideration is the effect of materials. The data (Table 3) indicate that significantly more spores are released from Eccobond than from methyl methacrylate. These test conditions are probably conservative for an actual spacecraft planetary impact in that Eccobond due to its hardness and brittleness characteristics may be a worse case. Impacting directly into

Table 3

Effect of impact material and velocity on percentage spore release

Impact velocity (m s^{-1})	168	457	945	1554[a]
Impact material				
Methyl methacrylate	0.09	0.31	0.06	0.02
Eccobond	0.16	0.63	0.43	0.19

[a] Data for methyl methacrylate at this velocity extrapolated to 10^4 initial inoculum level from 10^5 data.

stainless steel is also probably quite conservative because the probability of an impacting spacecraft encountering such a hard surface would be very low. Configuration of the spacecraft should also be a consideration in that the location of the materials will affect the impact. Considering the above factors, the data from these studies indicate that it is unlikely that the percentage release of internally contaminated materials would be greater than 1%.

The microbial release data may also be evaluated as log reductions. For example, a one log reduction would be a 90% decrease in the number of organisms released by impact when compared with the number of organisms present in the solid material before impact. The data reported in this paper suggest that a two log reduction results when the release data are applied to the buried microbial load. This can be significant when considered in terms of effect on the length of a lander sterilization cycle for encapsulated or buried organisms. The present concept is that flight hardware sterilization will be accomplished in a stepwise process [3]. The first step will be accomplished during the flight acceptance (FA) testing of the hardware. Each assembly level item will be subjected to a FA thermal test which will be sufficient to sterilize the interior of the item. The terminal thermal cycle will then be applied to the mated surface level in the assembled configuration. The relationship between the release of buried micro-organisms and the FA thermal treatment has been expressed by Schalkowsky [4] in the following equation:

$$P_B = M_B(O)\ 10^{-t_B/D_B}\ P_B(r)\ P_G$$

where P_B = allocation of mission contamination probability to buried load ($>10^{-6}$); $M_B(O)$ is the number of viable micro-organisms prior to FA heat cycle having a resistance to heat of $D_B \leqq 5$ hrs ($<10^{10}$); t_B is the number of hours at 125 °C during flight acceptance thermal cycle of components; D_B is the D value of buried organisms (5 hours); $P_B(r)$ is the probability that a buried organism will be released on the planet surface in viable state [1]; P_G is the probability that a released organism will cause proliferation of terrestrial biota on Mars (10^{-4}). If the probability that a buried organism will be released on a planet surface in a viable state were changed from unity to 1×10^{-2}, a reduction in the FA cycle of 10 hours would be effected.

4. Summary

The release of spores from internally contaminated solids was found to be dependent on velocity, material and impacting surface. The percentage of spores released was found to be higher from Eccobond than from methyl methacrylate except at the lowest velocity simulated and the percentage release from impacting stainless steel was higher at all velocities than from impacting sand. The percentage of spores released under all test conditions simulated was less than 1%. The release data were evaluated in terms of probability of release and the subsequent effect on sterilization processes for lander vehicles. The data obtained from these studies indicate that the probability of microbial release due to impact can probably be reduced from unity.

Acknowledgments

Major portions of this work were performed under JPL Contract 952511 and 952916 for the Jet Propulsion Laboratory, California Institute of Technology, as sponsored by the National Aeronautics and Space Administration under Contract NAS7-100.

References

[1] S. J. FRASER, R. L. OLSON and R. H. GREEN, Life Sciences and Space Research IX, 139 (1971).
[2] E. A. GUSTAN and R. L. OLSON, The Boeing Company, Seattle, Washington, Contract No. 952916, final report to Jet Propulsion Laboratory (June 1971).
[3] C. W. CRAVEN, J. A. STERN and G. F. ERVIN, Astronaut. Aeronaut. **6**, No. 8, p. 20 (August 1968).
[4] S. SCHALKOWSKY, Exotech Systems, Inc., personal communication (1971).

Life Sciences and Space Research X — Akademie-Verlag, Berlin 1972

EFFECTS OF AEOLIAN EROSION ON MICROBIAL RELEASE FROM SOLIDS

E. A. Gustan[a], R. L. Olson[a], D. M. Taylor[b] and R. H. Green[b]

[a] The Boeing Company, Seattle, Wash., USA
[b] Jet Propulsion Laboratory, California Institute of Technology, Pasadena, Calif., USA

Studies have shown that micro-organisms can become encapsulated in selected spacecraft solid materials and under specific conditions survive to arrive on planetary surfaces. This study was initiated to determine the percentage of spores that would be expected to be released from the interior of the solid materials by aeolian erosion on a planetary surface. The information obtained from the study can be used in calculations to determine the probability of microbial release in the total planetary quarantine probability equation. Methyl methacrylate and Eccobond disks were fabricated so that each disk contained approximately 4×10^4 *Bacillus subtilis* var. *niger* spores. The disks were placed in a specially designed sandblasting device and eroded. Exposure periods of 0.5, 2 and 24 hours were investigated using filtered air to accelerate the sand. A series of tests was also conducted for a 0.5 hour period using carbon dioxide. Examination of the erosion products showed that less than 1% of the spores originally contained in the solids was released by aeolian erosion.

1. Introduction

The basic planetary quarantine constraints for spacecraft have been established by the Committee on Space Research [1] and the National Aeronautics and Space Administration [2]. In order to comply with these constraints, the parameters for contaminating events have been identified and values have been assigned for the probability of contamination for each parameter. Originally, experimental evidence was not available for some of the parameters, therefore, very conservative values were chosen. One such value was assigned for the probability of the release of micro-organisms from the interior of solid materials. This value was taken as unity [3]. The importance of microbial release from solids was referred to by Sagan, Levinthal and Lederberg [4] and by Horowitz, Sharp and Davies [5] in their discussions of the Martian contamination problem. These investigators identified several mechanisms that could release organisms from solids, including such factors as a hard impact on a planet's surface and environmental erosion. Experimental data have previously been reported for assessing the probability of microbial release due to hard impact [6]. Data on microbial release from solids by aeolian erosion are presented in this paper.

2. Technical Discussion

Internally inoculated solid materials were eroded with sand in a specially designed sand blasting device. Due to laboratory time constraints, the erosion process was accelerated compared with predicted naturally occurring rates,

These aeolian erosion studies were designed to provide comparative information on the release of micro-organisms from two materials, at three erosion rates, and with two carrier gases. Six replicate samples were eroded for each test condition to provide statistically valid data.

The test materials were prepared by inoculating methyl methacrylate and Eccobond, an epoxy, with approximately 4×10^4 *Bacillus subtilis* var. *niger* spores per gram of material. The materials were then polymerized and fabricated into rods using a modification of the method reported by Angelotti et al. [7]. Disks were machined from these rods which measured 1.27 cm in diameter and 0.71 cm long. Each disk weighed approximately 1 gram. Aeolian erosion was accomplished by utilizing the Venturi principle in a specially designed sand blasting device (Fig. 1). A controlled flow of gas was metered into the apparatus creating a pressure differential in the Venturi tube which picked up sand from a reservoir. The sand was then accelerated through the nozzle and directed at a test disk positioned in the apparatus. A cyclone separator and 0.45 μm membrane filter were used to trap the dust that was formed and any spores that were released from the disk. The flow of gas was adjusted so as to erode 0.25 ± 0.05 grams from each test disk for erosion cycles of 0.5, 2 and 24 hours. Air and carbon dioxide were used as carrier gases for the 0.5 hour cycles, while only air was used for the 2 and 24 hour cycles.

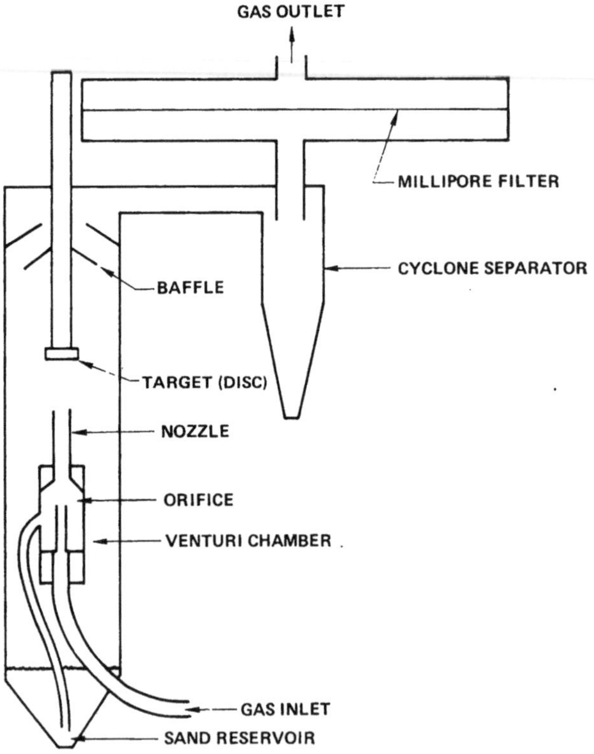

Fig. 1. Schematic diagram of sand blaster.

Following each erosion cycle, the sand blasting apparatus was dismantled and the contents analyzed for released viable spores. The remaining sand in the reservoir and the dust from the cyclone separator were placed in several petri dishes and mixed with trypticase soy agar (TSA). The apparatus was then rinsed with sterile distilled water to assure that all material had been removed. Any viable spores in the rinse water were recovered by membrane filtration. The eroded disk and the membrane filter from the apparatus were overlaid with TSA. All of the agar plates were incubated at 30 °C for 2 weeks. Each colony counted on the plates was recorded as one spore released from the disk.

Non-eroded disks were also assayed in order to determine the total spore concentration initially present within a disk. Randomly selected methyl methacrylate disks were dissolved in acetone and the viable spores recovered by membrane filtration. Since a suitable solvent is not available for Eccobond, these disks were assayed by a wet grinding procedure from which agar pour plates were prepared to determine the initial viable spore level. These Eccobond counts were adjusted with an experimentally determined constant to compensate for discrepancies inherent in the grinding technique [8].

3. Results and Discussion

Table 1 shows the gas flow rates measured in liters per minute as metered into the erosion device. The velocity of the gases through the orifice was approaching sonic velocity; however, the sand velocity was somewhat lower due to the size of the nozzle (4.7 mm inside diameter), air drag on the sand and a slight positive pressure in the apparatus due to the membrane filter.

Table 1

Gas flow rates

Erosion cycles (hours)	Flowrate (liter min^{-1})
0.5 – CO_2	105.2
0.5 – air	134.4
2 – air	119.3
24 – air	66.2

The percentages of spores released by aeolian erosion are shown in Table 2. A statistical analysis of the data shows no significant difference in the percentage

Table 2

Percentage of spores released by erosion based on 1×10^4 available spores

Erosion cycles (hours)	Percentage of spores released	
	Methyl methacrylate	Eccobond
0.5 – CO_2	0.11[a]	0.03
0.5 – air	0.02	0.002
2 – air	0.01	0.02
24 – air	0.02	0.01

[a]) mean of six replicates.

of spores released between the four erosion cycles or between the two materials. The data also show that an average percentage of 0.03 viable spores was released from the disks by erosion under all test conditions.

4. Summary

The aeolian erosion data for all test conditions show an average microbial release of 0.03% which is a log reduction of 3 in the initial microbial population. It is of significance to note that no apparent differences were observed for the release data between carrier gases, materials or erosion rates. The data obtained during this investigation indicate that the planetary quarantine probability of unity given for microbial release due to aeolian erosion should be re-evaluated and the probability reduced.

Acknowledgments

This work was performed under JPL Contract 952916 for the Jet Propulsion Laboratory, California Institute of Technology, as sponsored by the National Aeronautics and Space Administration under Contract NAS 7-100.

References

[1] COSPAR Information Bulletin No. 20, November 1964 (p. 24).
[2] L. B. HALL, in: COSPAR Technique Manual No. 4, 1968 (p. 3).
[3] L. B. HALL, personal communication (1971).
[4] C. SAGAN, E. C. LEVINTHAL and J. LEDERBERG, Science **159**, 1191 (1968).
[5] N. H. HOROWITZ, R. P. SHARP and R. W. DAVIES, Science **155**, 1501 (1967).
[6] S. J. FRASER, R. L. OLSON and R. H. GREEN, Life Sciences and Space Research IX, 139 (1971).
[7] R. ANGELOTTI et al., Appl. Microbiol., **16**, 735 (1968).
[8] E. A. GUSTAN and R. L. OLSON, Contract No. 952916, final report to Jet Propulsion Laboratory (June 1971). The Boeing Company, Seattle, Washington.

Life Sciences and Space Research X — Akademie-Verlag, Berlin 1972

SYNERGISTIC CHARACTERISTICS OF THERMORADIATION STERILIZATION

H. D. Sivinski and M. C. Reynolds

Planetary Quarantine Department, Sandia Laboratories, Albuquerque, New Mexico, USA

Temperatures required for thermal sterilization are known to degrade certain heat-sensitive components, materials and products. Simultaneous application of lower temperatures and low levels of gamma radiation produces a synergistic effect which can sterilize with fewer damaging side effects. With this technique sterilization times can be reduced significantly, for example, at 105 °C and a dose rate of 7.5 krad per hour, the D value for *B. subtilis* var. *niger* spores decreases from 4.5 hours for dry heat alone to 1.5 hours when both are applied simultaneously. Other spore systems also exhibit this synergetic inactivation characteristic; those studied include *B. stearothermophilus*, *B. pumilus* and some highly resistant soil spores. It was found that dose rate is a major factor to be considered in optimization of the synergism observed. Other effects studied include those of water activity and the use of a nitrogen atmosphere.

1. Introduction

In the search for increased spacecraft reliability it seems reasonable to study ways in which one can minimize the decrease in reliability caused by any given terminal sterilization cycle. One such study has resulted in the development of a process which combines dry heat and ionizing radiation in a way which results in greater microbial inactivation than the additive effects would imply. This process is referred to as thermoradiation in this paper.

The consequences of standard techniques, such as dry heat or ionizing radiation, frequently are unwanted side effects such as discoloration, embrittlement, change in electrical characteristics, and reduced reliability. Although dry heat is currently the approved sterilization process, ionizing radiation has also been considered for spacecraft sterilization [1]. The degradation resulting from the radiation doses necessary (2.5–5.0 Mrad) has caused rejection of radiation alone as a sterilant; the combination of radiation and dry heat however does show a great deal of promise. Our research [2] and the evidence of other investigators [3–7] in the biological effects of combined heat and gamma radiation has revealed complementary and sometime highly synergistic effects when these agents were simultaneously applied. Upon experimental confirmation of this synergism, we began a continuing research activity directed toward gaining insight into the synergistic mechanisms associated with bacterial inactivation and developing predictive mathematical models [8] to depict these relationships. We have investigated a number of controlling parameters in an effort to maximize the synergism so that sterilization can be accomplished at the lowest levels of stress to materials and components. Of primary concern was the development of the dose rate/temperature relationship at temperatures well below the present 125 °C sterilization requirement. With these relationships defined, a fullrange of options would be available from which to select the most appropriate temperature/dose combination for sterilization of a given material or product. These values were compared with the inactivation of heat alone, radiation alone, and radiation followed by heat treatment. In addition, we determined the effects of environmental moisture and of a nitrogen atmosphere on inactivation.

2. Materials and Methods

Spores of *Bacillus subtilis* var. *niger*, the working organism for NASA dry heat studies, were used in most of the experimentation for appraisal of relative inactivation of various treatments. Inactivation of *Bacillus pumilus* and highly resistant spores in soil has been in good agreement with the *B. subtilis* work.

A recirculating air oven placed at various positions within a ^{60}Co cell provided simultaneous thermal and gamma radiation at the desired dose rates between 3000 and 660000 rad per hour. Higher radiation rates of up to 10^{12} rad per hour were accomplished using a pulse X-ray machine with a mean photon energy of 1.6 MeV. Air within the oven was moisture conditioned to maintain the desired water activity level in the test organisms. Survivor curves were prepared from plate count data of successive sampling periods in a given environment. The D values (the time required for a 90% reduction of population) for comparison of destruction rates were calculated from linear regressions of the data. A series of experiments was performed to compare the sterilization effectiveness of dry heat alone, ionizing radiation alone, then various combinations of radiation and dry heat. The temperature range of interest was 125 °C down to 60 °C and radiation doses were limited to a level acceptable for typical spacecraft materials and components.

3. Experimental Results

Fig. 1, 2 and 3 are typical of the experimental results. Fig. 1 shows the results of our first studies which were done at 125 °C. Base-line studies are shown in the curves which are labeled as "radiation only at 23 °C" and the one using dry heat alone labeled "125 °C". A pre-irradiation treatment of 100 krad was given to one series of samples prior to dry heat only at 125 °C. The results are shown in the curve labeled "125 °C with radiation". The last curve shows the results of simultaneous irradiation at 50 krad/hr with heat at 125 °C. It is interesting to note that the curve representing simultaneous radiation and heat treatment at 125 °C is about three orders of magnitude (logs) below the 125 °C curve after two hours of treatment. If the environments were only additive, the simultaneous curve should be one log down from the dry heat curve since the radiation dose at this point was only 100 krads.

The results of a second series of experiments performed with a base temperature of 105 °C are shown in Fig. 2. The curves in Fig. 2 are similar to the 125 °C curves of Fig. 1 in that simultaneous application of heat and radiation is more effective than sequential application. The survivor curve representing simultaneous heat and radiation is six logs down from the dry heat curve with a gamma dose of less than 100 krad. One can conclude that a synergistic effect of approximately five logs has been experienced.

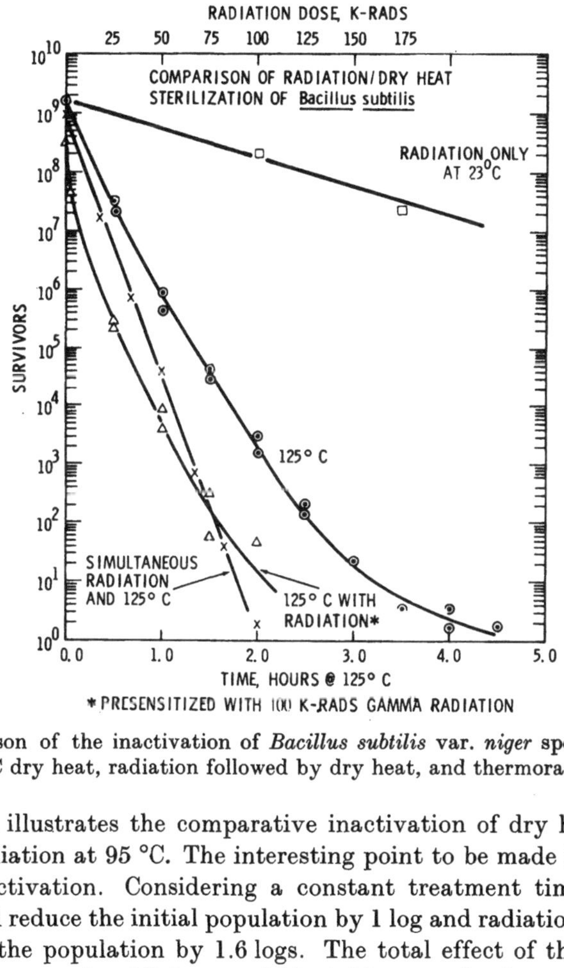

Fig. 1. Comparison of the inactivation of *Bacillus subtilis* var. *niger* spores with gamma radiation, 125 °C dry heat, radiation followed by dry heat, and thermoradiation.

Fig. 3 again illustrates the comparative inactivation of dry heat, radiation, and thermoradiation at 95 °C. The interesting point to be made is the degree of synergetic inactivation. Considering a constant treatment time of 12 hours, dry heat would reduce the initial population by 1 log and radiation at 11 krad/hr would reduce the population by 1.6 logs. The total effect of these two, if the inactivation were purely additive, would be 2.6 logs total reduction. A reduction of 5.2 logs or twice the additive effect was, however, observed.

From these data and others at differing conditions, it became obvious that the degree of synergism varied considerably. Consequently, it was desirable to augment the lower temperatures afforded by thermoradiation with optimization of any other parameters that could further reduce the stress of sterilization

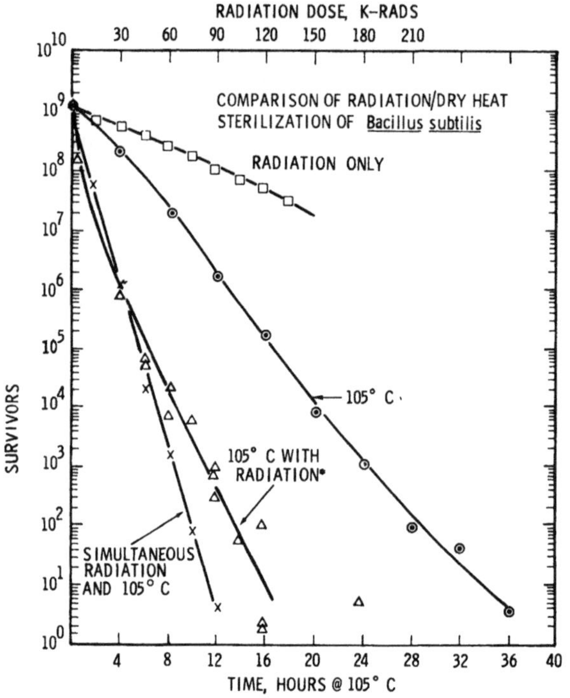

* pre-sensitized with 100 krads gamma radiation

Fig. 2. Comparison of the inactivation of *Bacillus subtilis* var. *niger* spores with gamma radiation, 105 °C dry heat, radiation followed by dry heat, and thermoradiation.

to materials. We investigated water activity effects on inactivation, radiation dose rate effects, population density effects, substrate effects, nitrogen effects and effects in plastics. Only the effects of water activity, dose rate, and nitrogen will be reviewed.

The effect of changing relative humidity within the laboratory is known to affect experimental results, particularly in dry studies. Using humidity control equipment to govern the relative humidity of the chamber, we performed a series of dry heat experiments at relative humidity extremes which most certainly cover the 35% RH level typical of our laboratory environments. The extremes chosen were 20–60% RH at room temperature. It is interesting to note that this spread in humidity represents a range of only 0.5–1.5% RH, respectively, when the air is heated in the oven to 100 °C. Dry heat survivor curves at 105 °C for *B. subtilis* var. *niger* under differing relative humidity conditions are compared in Fig. 4. There is indeed an effect on the inactivation by dry heat when the ambient conditions are varied from 20 to 60% RH. The 105 °C dry heat *D* value changes from 2.3 hours at 20% to 5.3 hours at 60% RH. The same experiments were next performed using thermoradiation (Fig. 5). It was interesting to find that within the range of dose rates used and at 105 °C, there seems to be little effect from changing ambient relative humidity during thermoradiation exposure. This is probably due to the different effect moisture has on

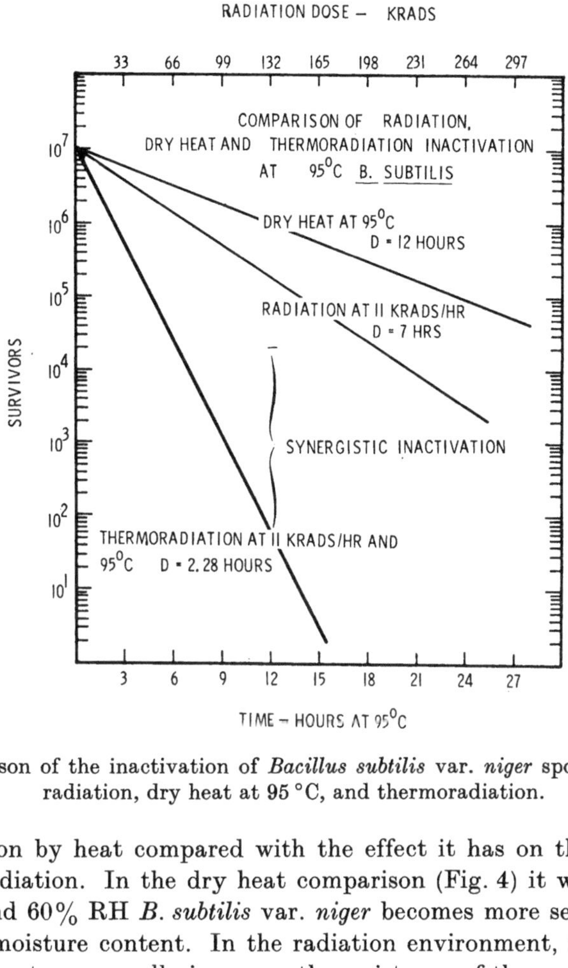

Fig. 3. Comparison of the inactivation of *Bacillus subtilis* var. *niger* spores with gamma radiation, dry heat at 95 °C, and thermoradiation.

the inactivation by heat compared with the effect it has on the inactivation by ionizing radiation. In the dry heat comparison (Fig. 4) it was shown that between 20 and 60% RH *B. subtilis* var. *niger* becomes more sensitive to heat at the lower moisture content. In the radiation environment, the removal of water from a system normally increases the resistance of the organisms concerned [9].

Since present plans for terminal dry heat sterilization of spacecraft include the use of dry nitrogen as a working gas, it is necessary to determine how thermoradiation inactivation is affected in an oxygen-free atmosphere. The oxygen dependence of radiation inactivation has been established by a number of investigators. For example, Proctor [10] found that an increase of 33% in the inactivation dose was required for *B. subtilis* spores in dry nitrogen compared with irradiation in air; Hollaender et al. [11] found even greater differences, that is, reduced radiation inactivation in nitrogen compared with air, for *E. coli*. No previous work was found that was applicable to nitrogen effects on simultaneous gamma radiation and dry heat inactivation. Base-line heat sterilization studies were done in nitrogen at 95 °C and thermoradiation experiments at 95 °C in N_2 were completed at dose rates of 4, 11, 38 and 85 krad/hr. The D values varied from about 1 hour at the 85 krad/hr rate to 3.4 hours at 4 krad/hr. The resulting

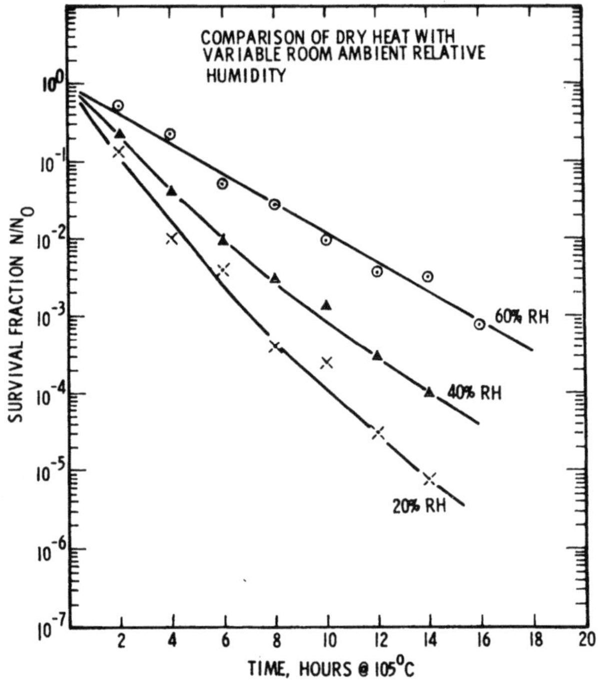

Fig. 4. Comparison of 105 °C dry heat inactivation of *Bacillus subtilis* var. *niger* spores with various levels of relative humidity.

curves are summarized in Fig. 6 and an interesting comparison is made in which the thermoradiation data in N_2 are compared with similar thermoradiation data in air. Although the D value in N_2 for high dose rates is generally somewhat higher than in air, at the optimal dose rate of 10 krad/hr there is essentially no difference.

The most interesting of the contributory factors whose effects were studied was that of dose rate. Fig. 7 shows the results of a series of experiments at 105 °C with dose rates ranging from 36 down to 2.6 krad/hr. The corresponding D values ranged from 40 minutes for the high rate to 2 hours at the lowest dose rate. The total dose for a given reduction in population was, however, significantly different. For example, if we consider a 4-log reduction in population, the high rate (36 krad/hr) has required a total dose of 96 krad where the lowest dose rate of 2.6 krad/hr has only required 21 krad. Thus, a rather striking dependence on dose rate is indicated. The difference in heating time between these two examples can only account for a small part of the dissimilar dose requirements. Another way to study the implications of Fig. 7 would be to consider a constant treatment time of 4 hours. Inactivation at the high rate (36 krad/hr) would result in a total dose of 144 krad during the 4-hour period with a 6-log reduction in population. The dose/log reduction would be 24 krad. The low dose-rate option (2.6 krad/hr) would result in a total dose of 10.4 krad during the same 4-hour period with a 2.5-log population reduction. In this latter

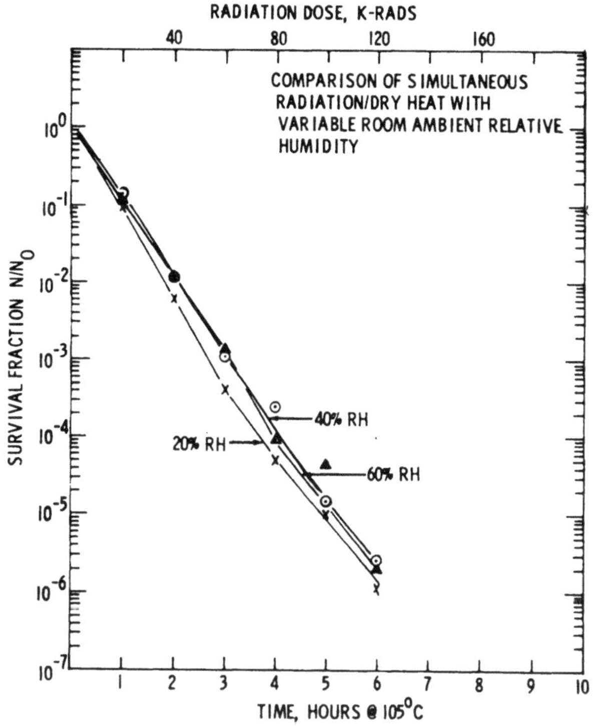

Fig. 5. Comparison of 105 °C thermoradiation inactivation of *Bacillus subtilis* var. *niger* spores with various levels of relative humidity.

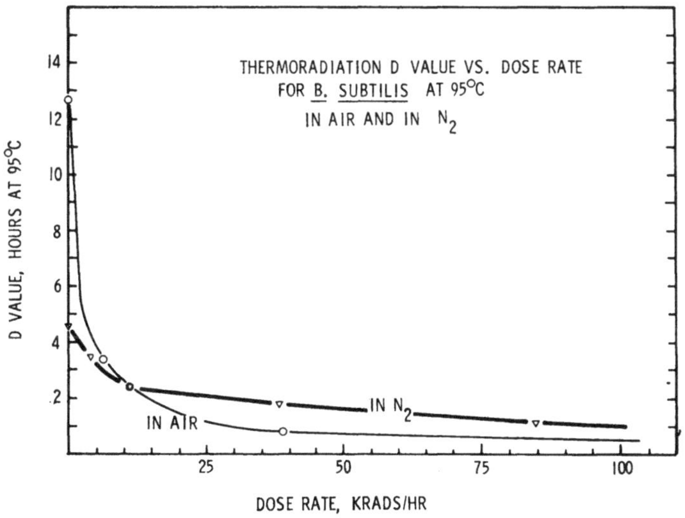

Fig. 6. Thermoradiation inactivation of *Bacillus subtilis* var. *niger* spores at 95 °C in air and in nitrogen.

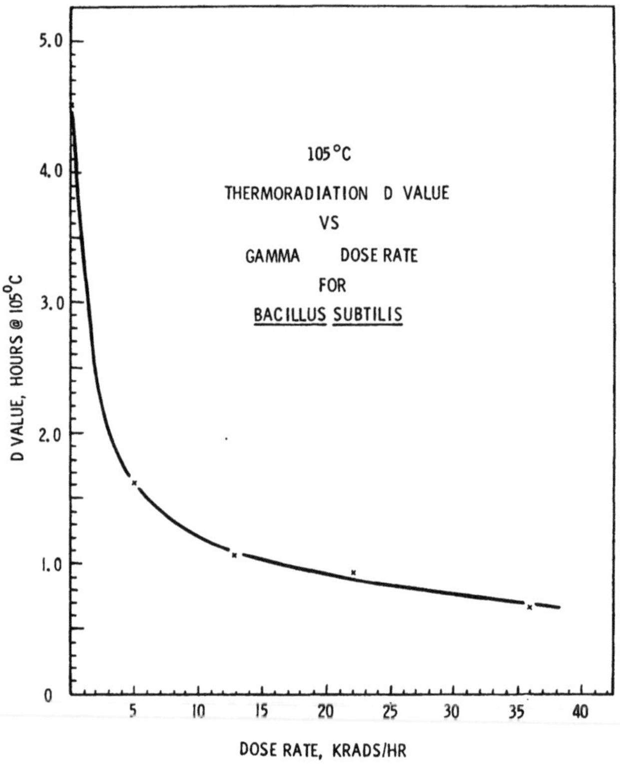

Fig. 7. Thermoradiation D values at 105 °C as a function of radiation dose rate for *Bacillus subtilis* var. *niger* spores.

case, the dose/log reduction would be 4.2 krads or about 1/6 the high rate requirement.

In another sense Fig. 7 also illustrates the trade-offs that can be made in a sterilization cycle. Beginning at the ordinate, the D value is that of dry heat at 105 °C or 4.5 hours. As low dose rate gamma radiation is added to the dry heat, the D value drops rapidly up to 10 or 15 krad/hr beyond which there is a marginal change in D value. For example, if one is interested in reducing a microbial population by 12 logs there are several options available. One option would be dry heat with a total sterilization time of 54 hours. A second option would be thermoradiation at 12 krad/hr with a D value of 1.1 hours, or 13 hours total sterilization time and a total dose of 156 krad. The third option for less radiation-sensitive material might be a dose rate of 36 krad/hr. At this dose rate the D value would be 0.7 hour, or 8.4 hours total sterilization time with a total dose of 302 krad. Accordingly, trade-offs in time at temperature and radiation dose can be made to adjust the sterilization cycle to the material to be sterilized. In this manner, some undesirable side effects of sterilization can be minimized. It is interesting to note that the D value vs dose rate curve for treatment in air at 95 °C (Fig. 6) has the same general shape as Fig. 7.

Thermoradiation seems most efficient when heat is combined with low dose rate gamma radiation. Using very high dose rate 1.6 MeV X-rays, however, has also been found to be effective but to a lesser degree. Fig. 8 shows a comparison of the inactivation of *B. subtilis* var. *niger* using X-ray, dry heat, and then thermoradiation (X-ray at 105 °C). As can be seen in this figure, after 6 hours of treatment the inactivation of dry heat and X-radiation in an additive sense would be $2^1/_2$ logs. Using thermoradiation, the inactivation is at least twice the sum of dry heat and X-ray inactivation or about 5 logs. Of interest, too, is the continued inverse effect of dose rate; for example, in Fig. 8 the thermo-

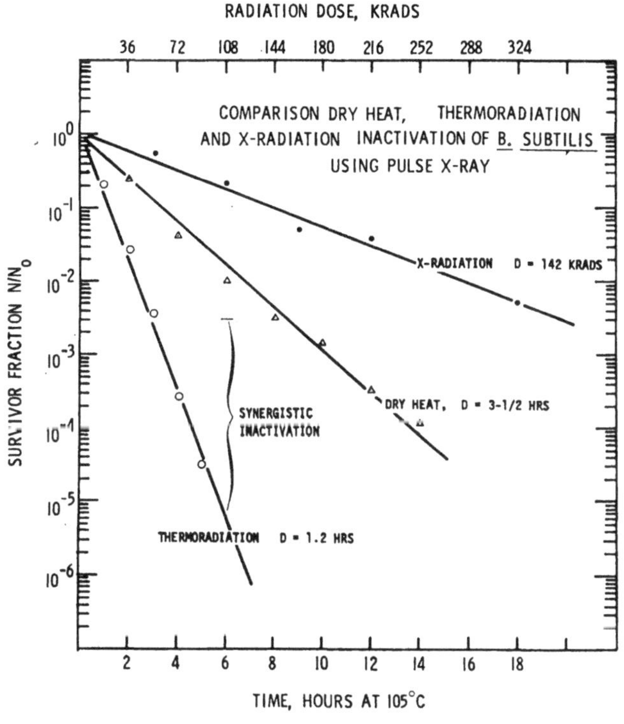

Fig. 8. Comparison of the inactivation of *Bacillus subtilis* var. *niger* spores with pulse X-radiation, dry heat, and thermoradiation at 105 °C utilizing pulse X-ray.

radiation *D* value at 105° with a single 18 krad pulse each hour was 1.2 hours. In Fig. 7, the *D* value using steady-state gamma radiation at 18 krad/hr was 1.0 hour. This dose-rate effect is better demonstrated in Fig. 9 where the inactivations by three dose rates at roomtemperature are compared. Inactivation at the high rate requires 2.2 times the total dose required at low dose rates.

Studies of some very resistant bacterial spores from soil have been initiated to determine the suitability of thermoradiation for sterilization of these naturally occurring populations. We have investigated the heat resistance of these spores at 105° and 125 °C, the radiation resistance at dose rates of 660 and 54 krad/hr, and the thermoradiation inactivation at both 105° and 125 °C. The dry heat

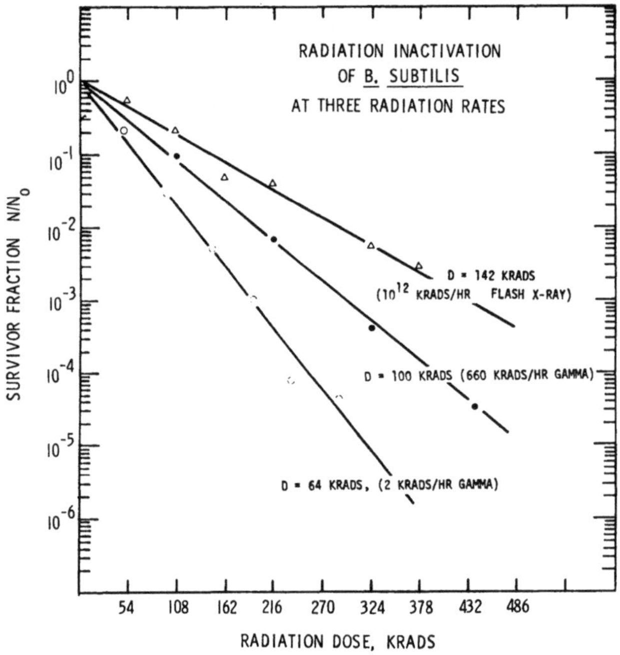

Fig. 9. Comparison of radiation inactivation of *Bacillus subtilis* var. *niger* spores at three widespread radiation dose rates.

destruction rate of the naturally occurring spores was found to be extremely slow at 105 °C; the D value for the resistant subpopulation (neglecting N_0) is 101.54 hours. This can be compared with a D value of $4^1/_2$ hours at this temperature for *B. subtilis* var. *niger*. In a subsequent experiment we determined the thermoradiation resistance of these naturally occurring spores by using dry heat at 105 °C combined with gamma radiation at 23 krad/hr. The D value derived from these data, again neglecting N_0, is 5.36 hours — $^1/_{20}$ of the dry heat D value. The radiation-resistance experiments were run at two dose rates, 660 and 54 krad/hr. We found a slight amount of dose rate sensitivity but an overall high degree of gamma radiation resistance. The D values obtained were 222 krad at 660 krad/hr and 205 krad at 54 krad/hr.

A number of experiments was performed at 125 °C to compare the thermoradiation resistance with the heat resistance of the naturally occurring spores. Fig. 10 is a comparison of dry heat, radiation, and thermoradiation inactivation of naturally occurring spores. In the dry heat experiment the D value derived from a least-squares fit of the resistant subpopulation was 29.45 hours. With thermoradiation of rate of 54 krad/hr and 125 °C, the D value was reduced to 1.04 hours. It is evident that substantial reductions in sterilization time are available when thermoradiation is used. A 30-fold reduction in time is available from a dry heat D value of nearly 30 hours to a thermoradiation D value of 1 hour. Fig. 10 shows the high degree of synergism experienced.

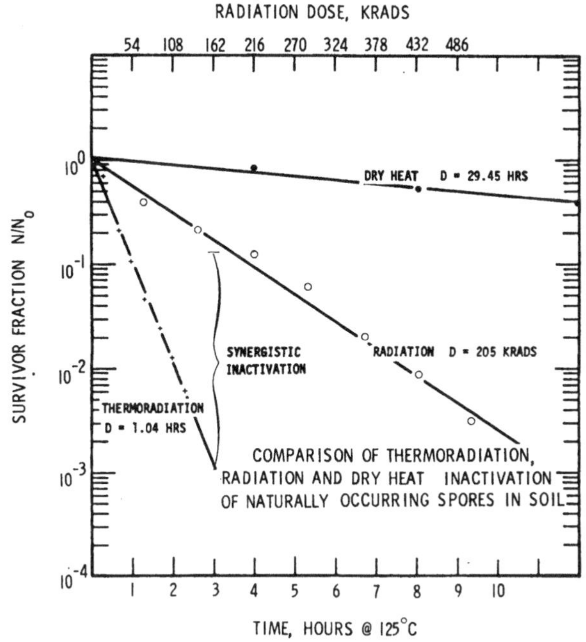

Fig. 10. Comparison of the inactivation of naturally occurring spores in soil with gamma radiation, dry heat at 105 °C and thermoradiation.

4. Conclusion

The striking dose rate dependence of radiation sterilization at temperatures around 105–125 °C suggests a great potential in low dose rate sterilization to minimize unwanted side effects. The longer exposure times associated with low dose rates fortunately can be reduced by the addition of heat. Thus thermoradiation at low temperatures and low levels of gamma radiation is indeed a means for sterilization whereby degradation of spacecraft, components, materials, and products can be minimized.

Acknowledgment

This work was conducted under Contract Number W-12, 853, Planetary Programs, Office of Space Science and Applications, NASA Headquarters, Washington, D.C.

References

[1] M. J. Barrett and W. C. Cooley, Exotech Inc. Report TR-012 (1966).
[2] M. C. Reynolds and D. M. Garst, Space Life Sci. **2**, 394 (1970).
[3] P. A. Buckley et al., Radiation Res. **40**, 26 (1969).

[4] K. Kaindl, in: Food Irradiation, International Atomic Energy Agency, Vienna 1966 (p. 701).
[5] B. Kan, S. A. Goldblith and B. E. Proctor, Food Res. **22**, 509 (1957).
[6] M. G. Koesterer, Develop. Ind. Microbiol. **6**, 268 (1964).
[7] E. L. Powers, R. B. Webb and C. F. Ehret, Radiation Res., Suppl. **2**, 94 (1960).
[8] V. L. Dugan, Space Life Sci. **4**, 498 (1971).
[9] G. Sykes, Disinfection and Sterilization, 2nd edn, J. B. Lippincott Co., Philadelphia 1967 (p. 162).
[10] B. E. Proctor et al., Radiation Res. **3**, 295 (1955).
[11] A. Hollaender, G. E. Stapleton and F. L. Martin, Nature, Lond. **167**, 103 (1951).

Effects of Weightlessness

Life Sciences and Space Research X — Akademie-Verlag, Berlin 1972

EFFECTS OF WEIGHTLESSNESS ON ASTRONAUTS — A SUMMARY

S. C. WHITE[a], C. A. BERRY[b] and R. R. HESSBERG[a]

[a] Office of Manned Spaceflight, Headquarters NASA, Washington, D. C., USA
[b] NASA Manned Spacecraft Center, Houston, Texas, USA

This paper reviews the adaptive changes observed in the United States astronauts during flight programs to this date. A series of postulates are offered as to what is happening in these adaptive events. A hypothesis is proposed as to the interrelationship of events observed in the body systems and functions involved. The importance of undertaking an extensive life sciences program, including an on-orbit phase of study as well as pre- and post-flight studies is discussed. Finally, the role the Skylab flight plays in the United States Space Program in achieving the future requirements for more extensive life sciences data is summarized.

1. Introduction

As we move into the second decade of manned space flight, it is appropriate that an assessment be made of the experience gained to date regarding man's capabilities to live and work in this new environment. Analysis of the information available reveals that much of the life sciences efforts during the first decade of flight were directed toward the reliable design, test of the spacecraft and the safe flight of man in the vehicle. This was done at the expense of an extensive research program to examine the basic mechanisms of man's accommodation to the space environment. Data obtained from the crew during flight were selected to provide a means of assessing crew status. Pre- and post-flight studies were directed to the assessment of whether man demonstrated the predicted physiological, psychological and behavioral effects expected or whether new and unexpected events may be occurring. Interestingly enough, the forecasts proved quite reliable. The degrees of change observed often varied from the predicted, but the body systems expected to change did show changes which generally followed in the direction predicted.

There were many complications in interpreting the flight data due to the occurrence of multiple non-space specific variables (i.e. periods of inadequate thermal control by the environmental systems within the spacecraft or within a space suit during an extravehicular operation; the presence of small crew space volume in the spacecraft forced a major degree of crew immobility; the artificial atmosphere selected for United States spacecraft provided a reduced total pressure of 258 mmHg and consisted of nearly 100% oxygen; crew time-line scheduling and system problems often caused inadequate food, fluid and electrolyte intake which was far below that programmed).

During the flights, man was frequently called upon to perform complex sensory and motor functions which demanded his best intellect and a maintenance of psychological stability. With rare exception, the crewman performed magnificently in meeting a wide variety of system problems. He proved a reliable multipurpose system backup and augmentative device. Analysis of the instances where he encountered difficulty in handling a problem raises questions as to whether his difficulty was due to limitations in his capacities or more related to a lack of understanding on all of our parts as to how to schedule his workloads for him, train him to meet some of the problems, or design the crew station to provide data to him in a manner permitting him to perform timely analysis of the problem and arrive at correct decisions for action.

2. Findings

In spite of the limitations noted in the Introduction, a considerable amount of information has been gained concerning man's ability to adapt to the space environment. As expected, the major new, unique space-induced stress to be encountered by man proved to be weightlessness. Space radiation, another space stress for man of special concern during the flights, has been of sufficiently low levels that doses received by the crewmen do not present a problem to date. The results presented concern the Gemini and Apollo missions.

2.1. Significant Medical Results — Gemini Missions

Table 1 gives a summary of the significant results from the Gemini missions.

A constant finding of a loss in body weight was recorded following each Gemini flight. This loss began to occur almost immediately after the onset of weightlessness and appeared to try to restabilize at a new point within the first day of flight. If there were additional problems that limited fluid intake or introduced thermal loads for the crews that caused sweating, the tendency for stabilization was not observed. Weight losses on the flights without complications were approximately 3–8% of the pre-flight body weight. After the shorter Gemini flights, this weight loss was promptly replaced within the initial 12–24 hours after landing. In the longest Gemini flight, the 14-day Gemini 7, a residual weight loss was also observed which extended beyond the first day after landing. This loss was recovered but required a longer period of time.

Table 1

Significant medical results — Gemini

Moderate body weight loss
Moderate cardiovascular deconditioning
Moderate loss of exercise capacity
Minimal loss of bone density
Minimal loss of calcium and muscle nitrogen
Moderate loss of red cell mass
High metabolic expenditure during extra vehicular activity

It is postulated that the initial prompt loss of weight may be attributed to a diuresis of body fluids considered excess after the earth gravity is removed. In the Gemini 7 case, it can be postulated that the mission was of sufficient length that the removal of gravity and the confined volumes of the spacecraft would cause the onset of muscle deconditioning. This could lead to loss of muscle mass which would require a longer recovery period. The metabolic assessment that was performed on Gemini 7 lends support to this postulate.

There was a consistent finding of moderate cardiovascular deconditioning coupled with loss of body fluids. The changes observed during post-flight testing included increased venous pooling in the gravity dependent areas, rapid increases in heart rates ranging up to 150–170 beats/minute when the crewmen were tested on a tilt table or required to perform tasks with moderate exertion, compression of the pulse pressure, and an occasional crewman approached syncope during the test periods. Recovery from these changes occurred spontaneously within 72 hours after landing. Lengthening the mission did not seem to affect the severity or the required recovery period. One could postulate that the cardiovascular picture may be due to changes in cardiac reflexes which control cardiac output, to reduction in total volume of circulating fluid and changes in levels of electrolytes. Obviously, since this is postulation, extensive study is required to define the mechanisms involved. If necessary the studies must also include methods of controlling the changes to a degree considered adequate for safe continuation and return to earth. Perhaps the use of the term cardiovascular deconditioning which has become popular in the Life Sciences literature is inappropriate since the changes observed appear to be normal adaptive events which should be expected in weightlessness. It should be noted that on no occasion was there any deterioration in crew performance during the flights. The cardiovascular changes only became apparent or offered a potential problem for the crew after the return to earth gravity. Man readapted to the earth environment during the 72 hour post-flight period.

There was moderate loss of exercise capacity after the longer Gemini missions. Testing was limited to comparisons of pre-and post-flight values. The test procedure challenged the crewman with a well controlled workload which was offered to him in a step-wise progression. Oxygen utilization, carbon dioxide production and length of time required for him to reach increased levels in heart rate were used as indicators of tolerance.

Through pre- and post-flight applications of an X-ray densitometry technique on selected bones, the *Os calcis* and the distal phalanx of the fifth finger, a reduction in the density of the bones as a function of duration of the space missions was observed. Owing to technical difficulties inherent in the measurement technique and the individual variations of the astronauts, the results of this test for calcium mobilization varied widely between crewmen on the same flight and between crews of different flights. On Gemini 7 a metabolic assessment study was performed; one of the parts of this study included estimation of the calcium and nitrogen balance. Although the test proved complex and encountered several operational problems during flight, there was evidence of increased calcium and muscle nitrogen loss. The time course and quantitative resultant of these occurrences were not determined. It is postulated that the lack of gravity reduces the stimulation of the bone needed to maintain calcium deposition in the bones and reduces muscle actions and their associated effects upon bone integrity, and the related changes in fluid and electrolyte balance are active in producing these changes. The importance of gaining definitive data on this matter for long-term space flight is obvious, and an extensive experimental effort is scheduled for Skylab to meet this need.

A moderate loss of red cell mass, as determined through pre- and post-flight studies, was also found in the Gemini flights. There has been much speculation as to the cause for this. The toxic effects due to the 100% oxygen atmosphere have been considered a prime suspect. A question as to whether the absence of nitrogen may be an important factor has also been considered. Interestingly enough, the loss of red cell mass appeared to be less on the 14-day Gemini 7 mission. From measurements following the Apollo 7 and 8 flights the regular finding of a loss of red cell mass appeared to stop. In Apollo 7 and 8 there had been a small residual of 3–5% nitrogen remaining in the spacecraft atmosphere throughout the mission. During the Apollo 9 mission, the command module was decompressed to permit an extravehicular operation and later repressurized with 100% oxygen. Again, the loss in red cell mass was observed. In Apollo 10, conditions returned to those of Apollo 7 and 8 where small residual amounts of nitrogen were maintained, and the increased loss in red cell mass was not observed. It was postulated following the Apollo 10 mission that the cell mass loss was related to the atmosphere either because of the toxicity of the 100% oxygen or because nitrogen was protective. Therefore, the test was discontinued until Apollo 14 since all spacecraft would carry residual nitrogen in their atmospheres. Future Apollo missions beginning with Apollo 15 may again have periods of 100% oxygen atmospheres since there may be extravehicular operations during the return from the moon. Therefore, the study of the red cell mass question was restarted on Apollo 14, in which residual nitrogen was maintained in the command module. Atmospheric conditions, therefore, were similar to those on Apollo 7, 8 and 10. Again, a moderate red cell mass loss was observed. Obviously, this opens the issue again and a study will be required to define and understand this problem.

The extravehicular operations were accompanied by an unexpected high metabolic expenditure, as calculated by oxygen utilization and the thermal loads which the environmental system was required to handle. As the Gemini program progressed, it became clear that this was not an inherent problem of space flight and weightless operation but a demonstration of our lack of understanding as to how to plan, train for and execute extravehicular actions in the weightless operation. By the close of the Gemini program, this problem was under control and metabolic loads, the thermal loads and fluid loss due to profuse sweating resulting from the workloads, approached the expected metabolic costs which ranged between 300 and 450 kcal/hour. By following heart rate the medical monitors on the ground were able to adjust workloads during the extravehicular operation to keep the astronaut from becoming overtaxed.

2.2. Significant Medical Results — Apollo Missions

In the Apollo missions many of the Gemini findings continued to be observed (Table 2), particularly those findings related to weight loss, cardiovascular deconditioning (adaptation), bone densitometry and loss of exercise capacity. The questions relating to the loss of red cell mass observed in Apollo have been discussed above and need no further elaboration here.

Table 2

Significant medical results — Apollo

No loss of red cell mass
Moderate cardiovascular deconditioning
Moderate loss of exercise capacity
Minimal loss of bone density
Moderate body weight loss
Low metabolic expenditure during lunar extra vehicular activity
In-flight motion sickness

There was a reduction in the energy expenditure required to perform lunar extravehicular activity. This represented further confidence that we understood how to plan and train for extravehicular work. It also reflects marked improvement in the extravehicular suit technology which permitted even more complex tasks to be performed than those in Gemini but at a much reduced energy cost for the men. Monitoring of the crew, through voice and heart rate, has also permitted valuable real-time evaluation of crew status which has been used to prevent lunar work periods from getting out of hand. The energy expenditure for lunar workloads, as calculated up through Apollo 14 which was the most ambitious of the crew work schedules to date, ranged between 200 and 300 kcal/hour. Some peak rates up to about 625 kcal/hour have been observed but these were of short duration and presented no problem for the crew.

With the beginning of Apollo flights, the United States astronauts encountered their first occurrences of motion sickness. The Apollo vehicle permitted the astronaut sufficient volume to permit him to move about and perform complex tasks at different stations within the cabin. The episodes of motion sickness occurred only during the initial five days. Most were triggered by a required rapid movement by the men. Symptoms ranged from stomach awareness to one incidence of vomiting. Symptoms spontaneously subsided and did not recur. One can postulate that the increased activity and mobility and the initial period of adaptation to the space environment may have all contributed to this new finding; further, one can postulate that after the astronauts once become adapted to the weightless environment, no further problem would be expected.

The last and most recent finding in the Apollo flights is the observation of the light flashes reported by the crews of all flights beginning with Apollo 11. There is no indication that the flashes are due to weightlessness effects; it is being included here only for completeness. It is thought that this phenomenon may be the result of high energy cosmic radiation striking the visual system of the crewmen. There is an active program to explore this: there will be attempts to correlate these events with tracks in nuclear emulsion dosimeters strategically located on the temples and forehead of one of the astronauts on the Apollo 16 mission. If the postulate as to the origin of the light flashes is correct, we are of the opinion that there is no hazard to the crew for the future lunar flights in Apollo or for the near earth Skylab orbital program. This would require, however, further intensive study before one was ready to commit for longer interplanetary flights of long duration outside the magnetosphere of the earth.

2.3. Some Effects of Null Gravity on Man — Current Working Hypotheses

In the survey above of the physiological changes observed during the Gemini and Apollo flight programs, various hypotheses were put forward as to what could be happening in the adaptive processes and in some of the interrelationships between measured changes. These are summarized in Fig. 1 to determine what is occurring within the total man. In this figure, the small arrows immediately adjacent to a function or element of the body are used to indicate an increase or decrease in the measured value of this function when pre- and post-flight measurements were compared on the same crewman. The larger or longer arrows which connect body functions or systems are used to show the forecast interrelationships of changes in one area with those in another.

Of special importance in this hypothesis is the forecast interrelationship of the cardiovascular deconditioning (adaptation) status with the fluid/elec-

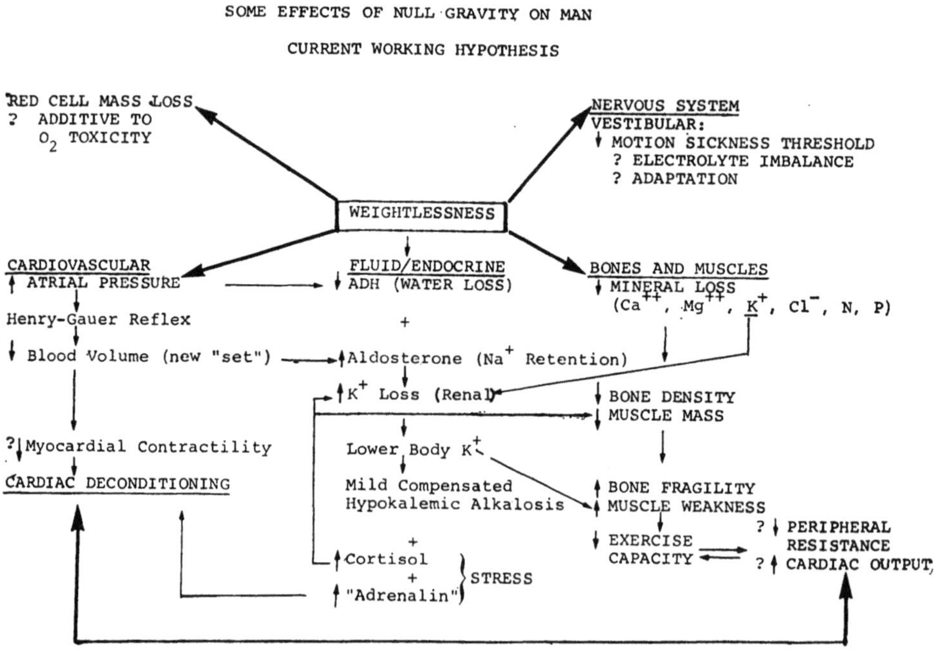

Fig. 1. Some effects of null gravity on man — current working hypothesis.

trolyte changes, the stress-produced hormone changes and the complex mineral shifts in the bones and muscles. The hypothesis proposes that these changes will be initiated upon the beginning of the exposure to weightlessness, move to new levels appropriate for the body to function in the weightless environment, then stabilize at these new levels. This stability, once achieved, would be expected to remain as long as there are no new serious stresses added. This hypothesis follows the course of events observed in man as he has been exposed to other hostile environments which did not exceed his total limits to adapt. If this hypothesis proves valid, operationally the crewman should be expected to achieve this new stabilized physiological condition after a period of transition. He should then be able to operate successfully in the weightless environment for longer periods of time. Only when he is rapidly returned to the earth gravity environment during entry and landing would he be forced to go into another re-adaptive sequence. Experience indicates that the crewmen have been able to make the on-orbit adaptation transition thus far without major problems if given a work schedule permitting a gentle transition for the first few days. On the other hand, the crewmen must be capable of meeting the dynamics of entry and the physical demands of the immediate post-landing period. Whereas the transition to the weightless stabilized position can afford several days, the entry and post-landing sequence is completed in a matter of minutes or hours. The entry and post-landing events, therefore, may set the program of on-orbit conditioning required to maintain the crewman in a condition sufficient for him to manage the return to the earth environment. If the hypothesis proves invalid, then progressive changes in physiological functions will be observed unless actions are taken to limit the change through the use of protective or supportive devices or procedures.

It is essential, therefore, for the life scientist teams to pursue programs that will answer these questions in a definitive way. If protective or supportive devices or procedures are required, it is important that these be designed based upon the understanding of what is happening to man in the space environment. The postulates proposed for explaining what is happening, the hypothesis which attempts to draw these together and the extensive experience gained by the life sciences field in the study of man as he has moved into other hostile environments, are being used to formulate the plans for future studies concerning long-term weightless operation.

The crew related findings observed during the Gemini and Apollo flights have led life scientists in the United States to stress the urgent need to accomplish an extensive medical-bioscience study program to clarify the questions of the adaptation of man to the space environment. This program must include an on-orbit phase as well as the presently used pre-flight and post-flight phases. Only through such efforts can the required definitive criteria for vehicle design and planning of longer durations of manned missions be acquired. From the reports received thus far from the Salyut–Soyuz 11 flight, it appears that similar needs have been stressed by the Soviet life scientist teams.

The Skylab program, scheduled to fly in 1973, incorporates these considerations in the experiment planning. It is the first attempt within the United

States to undertake such a broad experimental program. The initial manned flight of Skylab will aim at achieving a 28-day duration. If this is successful and the data resulting from the first mission indicate that it is safe to proceed, the Skylab revisits by new crews will attempt to operate for 56 days on each of the last two revisits. Fig. 2 summarizes the life sciences program presently

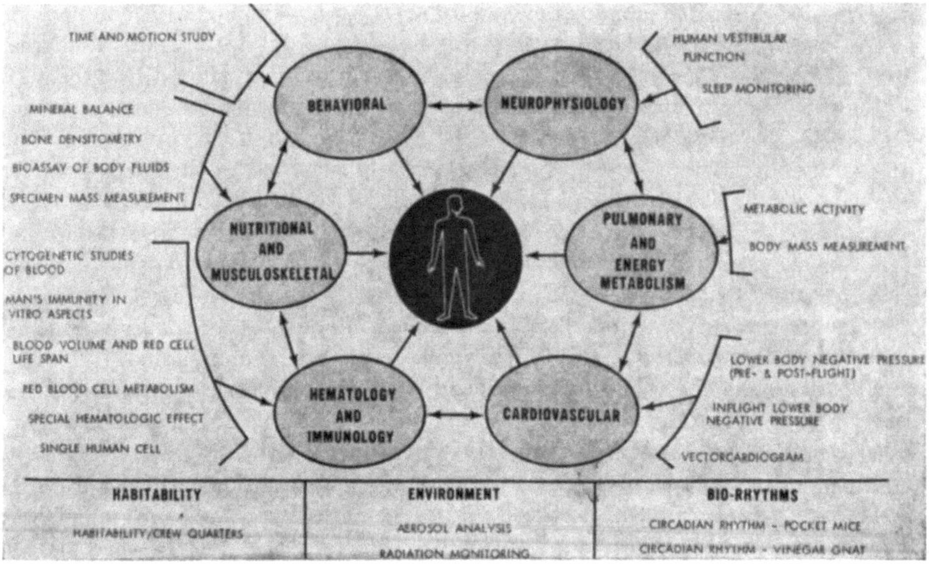

Fig. 2. Life sciences experiment relationships.

proposed for Skylab. It will be seen that most of the body systems or functions that have shown change in the Gemini and Apollo flights are included in the Skylab investigations. In addition, new areas have been added where it is predicted that the Skylab mission durations may produce changes of concern (the study of sleep analysis is an example), an attempt will be made to assess man's capacity for performance more precisely, and, an important addition, an in-flight phase will be added for many of the experiments previously done only as pre- and post-flight studies during Gemini and Apollo.

The objectives of each of the major segments of the life sciences experiments scheduled for Skylab are summarized in Table 3. It should be noted that the Skylab offers only the initial opportunity to collect such important data for future programs and for understanding man's ability to adapt. It cannot be considered definitive.

An extensive ground-based data collection program is planned for Skylab; this will develop the control and reference data base to be used for interpreting the flight results. It is also hoped that the Skylab experiments program will provide the basis for establishing a better understanding of the relationship of ground simulation to the actual flight events. If successful, it could offer future programs the option to use ground-based studies more efficiently and

Table 3

Skylab medical experiments program — Major areas of interest

M070 Nutritional and musculoskeletal function
Determine extent of skeletal and muscular alterations; evaluate mineral, water, electrolyte, and hormone changes; assess nutritive requirements.

M090 Cardiovascular function
Assess effects of space flight on circulatory system.

M110 Hematology and immunology
Determine space flight effects on physiology of formed blood elements, body fluid compartments, hemostatic mechanisms and selected aspects of immunity.

M130 Neurophysiology
Evaluate effects of the space environment upon the nervous system of man.

M150 Behavioral effects
Assess man's functional efficiency in completing operational and scientific work during long duration space flight.

M170 Pulmonary function and energy metabolism
Assess effects of spacecraft environment on crew's respiration, body mass and composition, and energy costs of physical activity.

with greater confidence. Early reports from the Salyut–Soyuz 11 flight indicate that a program closely related to that proposed for Skylab is now being conducted by the Soviet life sciences team.

References

Most of the data related to the findings reported here have been published or presented at international meetings or in recognized scientific journals as the data were being obtained. The best sources for reports on United States results are contained in the *Journal of the Aerospace Medical Association, Proceedings of the International Astronautics Federation,* the *Proceedings* of the three International Symposia on "Basic Problems of Man in the Space Environment" sponsored through the International Academy of Astronautics, and the Mercury, Gemini and Apollo Program Reports published by NASA.

Life Sciences and Space Research X — Akademie-Verlag, Berlin 1972

EFFECTS OF AN 18-DAY FLIGHT ON THE HUMAN BODY

A. D. Yegorov, L. I. Kakurin and Yu. G. Nefyodov

Institute of Medical and Biological Problems, Moscow, USSR

During their flight on board the Soyuz 9 A. G. Nikolayev and V. I. Sevastyanov adapted to weightlessness by the 3rd–4th day and stabilized their physiological functions by the end of the mission. During flight both cosmonauts maintained normal performance, sleep and appetite; they readily developed a new stereotype of movements and exhibited no noticeable increase of the circulatory function in response to a standard physical load.
In contrast to the effects of shorter term flights, this mission caused unusual and distressing feelings in the crew members aggravated by distinct changes in the major physiological systems during the first day of recovery. In the immediate post-flight hours the transition from the recumbent to the sitting position brought about circulation disorders; 24 hours later the cosmonauts still walked with uncertainty and kept the erect position at rest on account of a significant elevation of their centre of gravity. Weight losses, shifts of water and mineral metabolism, bone tissue demineralization and symptoms of orthostatic intolerance observed in this flight were similar to those resulting from earlier short-term missions. Of importance was a dysbacteriotic change in the skin and nasal microflora. Physiological changes in the Soyuz 9 crew members were functional and reversible, being on the whole in agreement with predicted effects.
These results call for the development of specific measures facilitating the post-flight adaptation of space pilots in view of future long duration space flights.

The main purpose of the in-flight medical investigations was a further study of changes in various physiological systems of man during adaptation to weightlessness and readaptation to 1 *g* gravity on return on earth.

The atmospheric parameters in the space cabin remained within the assigned limits, being similar to the normal terrestrial atmosphere. The radiation environment was favourable and the dose absorbed was not more than 0.4 rad. The menu consisted of natural canned foods, with caloric value averaging 2700 cal/day. Water consumption was 1.6–1.8 litres/day on the average. During flight the cosmonauts regularly — twice every day — performed physical exercises which followed a three-day cycle. Each cycle consisted of specific exercises that were intended to maintain muscular strength and speed (first day), muscular endurance (second day) and general endurance and stamina (third day). The load on all muscle groups was distributed uniformly with higher priority given to simulated walking, running and jumping on the spot.

Throughout the whole flight the cosmonauts estimated their health condition as excellent or good. A detailed analysis of their personal reports revealed interesting data. Both cosmonauts pointed out that occasionally they had feelings very similar to those experienced on earth during exposure to Coriolis accelerations. In flight this occurred when the spacecraft was spinning and the cosmonauts jerked their body or head, their feet being restrained. While floating freely in the cabin, their eyes closed, the cosmonauts rapidly disoriented in relation to their position within the cabin. In the weightless state the cosmonauts had the sensation of blood rushing to the head which brought about a flushed and hyperemic face. The strength of this sensation tended to decline after the first day and to disappear later.

The cosmonauts reported interesting peculiarities concerning their motor performance in weightlessness. In early flight they were at a loss to estimate correctly muscular efforts required for various movements and these were occasionally inadequate. However, by the third or fourth day the cosmonauts restored the necessary precision of movement which indicated an establishment of a new motor stereotype.

Voice communications, television broadcasts and personal reports suggested that they behaved properly and adequately to their individual features and operational situations. The work capacity of the crew members remained high throughout the flight. Nevertheless, beginning with the 12th–13th day the cosmonauts noticed that after an intensive working day they developed a feeling of fatigue. Their appetite was good and thirst somewhat lowered. Urination and defecation were normal. Following several flight days, sleep was adequate, averaging 7–9 hours; the cosmonauts characterized their sleep as sound and refreshing.

According to the telemetric data, during the third or fourth orbits the pulse rate approximated the values recorded a month before the flight and was stabilized at a level which was lower than pre-flight. Later during the last 6 days the pulse rate tended to increase. During or immediately after physical exercise both crew members showed an increase of the pulse rate which remained within 100 beats/min. The maximal and minimal arterial pressure at rest as measured by the cosmonauts themselves varied within 128–120 and 70–80 mm Hg, respectively.

Post-flight clinical and physiological examinations revealed noticeable changes in the function of several systems as well as symptoms of asthenia and fatigue. The motor and cardiovascular systems displayed most distinct variations. Immediately upon landing the cosmonauts reported their general weakness. When walking or doing light work, they displayed an unusual increase of the pulse rate. On return to earth the cosmonauts felt their heads, limbs and objects they used heavy as if they were under an acceleration stress of 2–2.5 *g*. During the first three post-flight hours the cosmonauts found it difficult to keep an upright posture and had to be assisted while walking; therefore, they preferred to remain recumbent. The next day after return their walking was still inadequate and the erect position was maintained with great effort. Stabilographical examinations of both cosmonauts showed a marked increase in the amplitude of displacements of the total body mass centre.

The results of the post-flight medical observations of the crew members are of great interest. As was expected, their orthostatic tolerance declined. Shortly after landing the cosmonauts found it difficult to change their body positon from horizontal to vertical. The change of posture brought about a significant increase of the pulse rate. V. I. Sevestyanov performed the first tilt test two days and A. G. Nikolayev three days after landing. Both cosmonauts exhibited a distinct post-flight decline of orthostatic tolerance (Table 1). Medical examinations of the Soyuz 9 crew members performed 11 days post-flight indicated that their orthostatic tolerance was close to the pre-flight level.

Table 1

Circulation changes during passive orthostatic tests

	Examination time	Pulse rate (beats/min)		Stroke volume (ml)			Cardiac output l/min			Oxygen pulse (ml O_2/beat)		
		RP	EP	RP	EP	%	RP	EP	%	RP	EP	%
A. G. Nikolayev	Pre-flight	73	97	67.2	47.4	−29.5	4.91	4.60	− 6.3	3.21	2.62	−48.4
	Post-flight											
	3rd day	71	103	94.5	50.9	−46.1	6.72	5.26	−21.7	3.36	2.45	−27.1
	6th day	79	117	63.2	34.4	−45.1	4.98	4.02	−19.5	2.83	2.22	−21.6
	10th day	71	100	70.8	—	—	5.02	—	—	2.99	2.45	−18.1
V. I. Sevastyanov	Pre-flight	68	93	78.8	52.9	−32.9	5.36	4.92	− 9.2	3.63	2.83	−22.0
	Post-flight											
	2nd day	62	125	91.5	47.5	−48.1	5.67	5.92	4.8	4.11	2.78	−32.4
	6th day	77	102	92.3	54.2	−41.3	7.10	5.53	−22.1	3.29	2.50	−24.0
	10th day	73	93	78.4	53.6	−31.6	5.72	3.98	−22.9	3.28	3.06	− 6.7

RP, recumbent position; EP, erect position.

During an 18-day space flight A. G. Nikolayev and V. I. Sevastyanov lost 2.7 and 4.0 kg respectively. Thus, no significant difference in the body weight losses during short- and long-term space missions was noted. Within three days the cosmonauts restored their body weight only partially: A. G. Nikolayev by 50% and V. I. Sevastyanov by 38%. However, even by the tenth post-flight day, their body weight did not reach the pre-flight value. Water losses were 1.4 kg in A. G. Nikolayev and 1.7 kg V. I. Sevastyanov (52% and 42.5% of the total body weight loss, respectively). Insignificant variations in the hemoconcentration parameters (hydrophylic property, hematocrit, clotting factor) suggest that the circulating blood volume remained within the pre-flight limits. The change of muscle tone, leg volume and increased excretion of nitrogen, potassium and sulphur are indicative of relative muscle atrophy.

Bone density was measured by means of radiophotometry and ultrasonic radiodensitometry. The data obtained are presented in Table 2. The recovery processes in all bone areas tested developed in one direction: after the 18-day flight, the decline of bone density was similar to that found in the cosmonauts who made shorter space flights.

Readaptation to 1 *g* gravity was characterized by a number of functional disturbances, circulation disorders and changes in the static and dynamic maintenance of the erect posture being most significant. The latter phenomena

Table 2

Percentage changes in the optical density of bone tissue compared with base-line values

Area examined	Nikolayev		Sevastyanov	
	2nd day	22nd day	2nd day	22nd day
1st phalanx of the hand				
3rd finger	−5	−2.5	−5.0	−5.0
4th finger	−3.1	±0	−4.3	−1.4
5th finger	−4.7	−1.6	−8.9	−4.4
Central segment of heel bone	−8.5	−4.5	−9.6	−3.4

were very distinct and particularly distressing. There is no doubt that readaptation on return to earth involves greater difficulty than adaptation to weightlessness. Thus, the problem of readaptation deserves particular attention. The results of the medical examinations give evidence that, having in view extended space missions, specific investigations and development of prophylactic measures facilitating the human body readaptation to 1 g gravity are required.

Life Sciences and Space Research X — Akademie-Verlag, Berlin 1972

FUNCTIONAL INSUFFICIENCY OF THE NEUROMUSCULAR SYSTEM CAUSED BY WEIGHTLESSNESS AND HYPOKINESIA

L. I. Kakurin, M. A. Cherepakhin, A. S. Ushakov and Yu. A. Senkevich

Institute of Medical and Biological Problems, Moscow, USSR

The results of study of the crew members of the spaceships Soyuz are described, and the effects of weightlessness on reflex excitability, muscular tone and muscle contractibility discussed. A certain decrease in postural muscular tone and strength, increase in reflex excitability at rest and increase in bioelectric activity of muscles at work has been found in the cosmonauts after their stay in a weightless environment. The circumference of the lower extremities decreased.

Data on prolonged exposure of human neuromuscular apparatus to weightlessness is scanty at present. Simulation of some aspects of weightlessness by means of hypokinesia showed changes similar to those found in the crewmen following their space flights (decrease in strength and tone of postural muscles, decrease in optic density of bone structures, body dehydration etc. [1]).

According to the data available [2], periods of up to 14 days in a weightless environment caused neither loss of coordination of movement nor muscular atrophy; at the same time certain alterations in some postural responses were revealed. The results of examination of the crew members of the spaceships Soyuz are presented in this paper.

To evaluate the state of the neuromuscular system of the cosmonauts before the flight and during readaptation to the ground environment after the flight the reflex excitability, muscle tone and muscle contractibility have been studied. The reflex excitability was measured by registering the bioelectric activity in the muscles concerned in the knee-jerk reflex. Muscle tone was measured by the method of Sirmai.

The circumferences of the extremities were measured with a steel tape at a fixed level which was checked with a measuring bar. The tension of the tape was kept steady by a standard weight.

No cosmonauts showed any alterations in the function of movement after being in the weightless environment, and no cosmonauts had any complaints after their short-term flights. Following the 18-day space flight, however, there appeared body weakness, muscular pains in legs and back, instability in the vertical position. The cosmonauts tended to sit down during the examination. On the first day of the examination it was noted that the amplitudes of the biopotentials of the knee-jerk reflex muscles had increased twice in A. G. Nikolayev and three times in V. I. Sevastyanov compared with their pre-flight values. The same trend was also noted in the case of short-term flights. The cosmonauts had a sensation of pain on tapping of the tendons with a neurological hammer. The intensity of pain decreased with each subsequent examination and disappeared on the seventh day.

The reflexes were also found to be greater on one side (the right); in Nikolayev this difference disappeared by the 11th day following the flight, while in Sevastyanov it was still pronounced at this time. On the 36th day after the flight the reflexes were provoked with difficulty on the left as well as on the right sides in both the cosmonauts (Table 1).

Table 1

Dynamics of bioelectric activity of muscles taking part in the knee-jerk reflex (mV)

Periods of observation	A. G. Nikolayev		V. I. Sevastyanov	
	Right leg	Left leg	Right leg	Left leg
Before flight	60.2		70.0	
Days after flight				
2	145.9	153.8	264.0	162.0
3	88.3	78.7	—	—
4	84.5	55.7	142.08	51.8
7	74.8	44.5	103.2	49.8
11	80.0	82.0	66.0	24.0
36	24.0	18.0	20.0	18.0

Muscular tone in the lower extremities was found to be attenuated after the flight in both the cosmonauts but the tone of the brachial muscle even exceeded the original level. The tone practically recovered in both the men by the 11th day. The strength of the hand flexors after the flight did not change. The strength of torso extensors became 40 kg less in A. G. Nikolayev and 65 kg less in V. I. Sevastyanov, compared with the original level; strength was recovered by the 11th day. The cosmonauts described sensations of pain in leg and back muscles (Table 2).

Table 2

Muscle tone as determined by the method of Sirmai (in relative units)

Periods of observation	A. G. Nikolayev			V. I. Sevastyanov		
	(1)	(2)	(3)	(1)	(2)	(3)
Before flight	102	56	69	105	78	78
Days after flight						
2	92	50*	75	92	66	60
3	92	53	—	—	—	—
4	94	51	—	96	68	—
7	96	54	74	102	70	76
11	101	54	75	103	68	77
36	107	56	74	102	71	78

(1) m. tibialis anterior; (2) m. quadriceps femoris; (3) m. biceps brachii.

* Postflight tibial muscle tone in A. G. Nikolayev was below the minimum (50) of the scales.

2 cm decrease in circumference of the legs and 3 cm decrease in that of the thighs were measured; by the 11th post-flight day these extremity circumferences corresponded to the original values (Table 3).

Table 3

Circumference of the extremities (cm)

Periods of observation	A. G. Nikolayev leg	thigh	arm	V. I. Sevastyanov leg	thigh	arm
Before flight	34.5	49.1	26.7	35.2	54.8	26.7
Days after flight						
2	34.1	47.0	26.8	33.2	51.6	26.3
3	33.9	47.8	—	—	—	—
4	34.0	48.0	—	33.9	50.5	—
7	34.4	48.0	27.1	34.1	51.7	26.6
11	34.7	49.0	26.7	34.9	51.9	26.9
36	35.0	50.7	28.4	34.8	54.1	26.4

The results obtained suggest that weightlessness is the main cause of decrease in muscle tone, since the need for active posture support and hence tension of the corresponding muscles is eliminated. The alterations concerned mainly those groups of muscles for which the burden is most decreased; thus, the brachial muscle tone was found to be even slightly higher than it was before the flight, while muscles of the leg and the thigh became 13 relative units less. Taking into account that the limits of fluctuations in tone for one man do not exceed 20–25 units, the decrease noted should be considered important.

Decrease in circumference of lower extremities as well as the body weight loss are most probably related to muscular atrophy. This agrees with the results of biochemical study performed after the flight which point to disturbances in nitrogen and calcium balances.

Soon after the short-term space flights signs of alterations in the nervous system were already revealed by increased amplitudes of the tendon reflexes at rest and a more prominent decrease in the level of reflex excitability in response to the post-flight physical load. The results of the 18-day flight showed increased reflex excitability of the neuromuscular system, asymmetry in tendon reflexes, and pain at the points of tapping with the neurological hammer; all these are signs of deterioration in the functional condition of the nervous system caused by sharp fluctuations in afferentation occurring at various stages of the flight, as well as at the periods of readaptation to the ground.

It is evident that an integral physiological index such as the neuromuscular tone, which serves as a vital regulatory mechanism for body functions, undergoes a certain rebuilding under the effects of space factors, in particular weightlessness, which is in many respects a determinant for the character of the readaptation process. This is confirmed by the results of our experiments, where the animals (rats) after prolonged confinement showed a decrease in activity of brain and cardiac cholinergic and adrenergic structures. Space flights of longer duration would probably complicate the adaptation to gravity, and great efforts would have to be made to prevent unfavourable consequences.

We assume, therefore, that active methods for the prevention of neuromuscular disturbances should use physical training to strengthen postural responses including standing, walking and support. Impact stress produced by the cosmonaut himself in jumps using shock absorbers would seem valuable in prevention of muscular and other disturbances. Such training would be similar in its effect to training in the gravitational field.

References

[1] O. G. GAZENKO and B. S. ALYAKRINSKY, Vestn. Acad. Sci. USSR, **11**, 40 (1970).
[2] C. A. BERRY, J. Am. Med. Ass. **201**, No. 4, p. 86 (1967).

Life Sciences and Space Research X — Akademie-Verlag, Berlin 1972

THE EFFECT OF EXERCISE ON THE PREVENTION OF CARDIOVASCULAR DECONDITIONING DURING PROLONGED IMMOBILIZATION

J. H. Triebwasser and M. C. Lancaster

US Air Force School of Aerospace Medicine, Brooks Air Force Base, Texas, USA

Much study has been given to the etiology of orthostatic hypotension that results from weightlessness and prolonged hypodynamia. The absence of gravity in some yet incompletely defined manner alters the cardiovascular integrity under stress. An ergometer was developed at the USAF School of Aerospace Medicine which applies force parallel to the long axis of the body similar to gravity during supine exercise. Sixteen normal men were studied over a sixteen-week period including five weeks of strict bed rest. During the first five weeks, control period, and the last six weeks, recovery, all sixteen expended 600 kcal of energy daily on the body ergometer. Eight of the subjects continued to exercise during the bed rest phase and the remainder served as their controls. Metabolic balance and blood volume determinations were performed throughout the study. Maximum treadmill exercise stress testing including measurements of oxygen consumption were obtained weekly during the control and recovery periods. Lower body negative pressure stress testing was performed every other week during the control and recovery phases and once during the fifth week of bed rest. Regional blood flow in the upper extremities was determined with a mercury in rubber strain gauge plethysmograph during the test. Plasma renin levels were measured before and during the last minute of negative pressure.

Twenty-four hours after bed rest, the subjects that did not exercise during immobilization had a 27% reduction in their mean maximum oxygen consumption. This decrement was 13.6 milliliters per kilogram or 876 milliliters ($P < 0.001$). No such decrement was observed for the exercise group. Twenty-four hours after bed rest, peak heart rates occurred significantly earlier in the non-exercisers than before they had bed rest. In addition these peak heart rates occurred considerably earlier than did those for the exercise subjects. All sixteen subjects demonstrated similar orthostatic intolerance when one uses the heart rate responses to lower body negative pressure following five weeks of bed rest. There were six episodes of syncope; five occurred in those individuals that exercised during immobilization. There was no decrement at any time in the plasma renin response to venous pooling in the lower extremities. There is some evidence to suggest that those who did not exercise during bed rest had an even greater increment in their plasma renin during lower body negative pressure stress testing.

These observations will be presented in more detail and correlated with the results of other metabolic studies including blood volume determinations in a paper in the *Journal of Applied Physiology*.

Life Sciences and Space Research X — Akademie-Verlag, Berlin 1972

STUDIES ON WEIGHTLESSNESS IN A PRIMATE IN THE BIOSATELLITE 3 EXPERIMENT

W. R. Adey

Space Biology Laboratory, Brain Research Institute, University of California, Los Angeles, USA

In June 1969 a male *Macaca nemestrina* (pigtail macaque) was flown in earth orbit for 8.8 days in NASA Biosatellite 3. The experiment examined in detail central nervous and cardiovascular functions, and included pre- and post-flight whole body metabolic assessment, in-flight urine analysis, and pre- and post-flight bone density measurements. Although the sleep/wake cycle was 24 hr, a phase angle lag of 2 hr from the imposed night/day mode occurred. A definite desynchronosis occurred, with rhythms longer than 24 hr in pCO_2, brain and body temperature and heart rate, although arterial blood pressure remained at 24 hr. Sleep states were remarkably fragmented and unusually brief in duration. Vestibular and ocular disturbances were evident. These changes began concurrently with onset of weightlessness and were not secondary to altered fluid balance or body temperature. Sleep patterns lie between those of normal man and man with high cervical cord transection. There was an immediate and sustained increase in central venous pressure in weightlessness and this is considered to have initiated a Henry–Gauer reflex which initially maintained a high urine volume. This, coupled with a high evaporative fluid loss, produced an early dehydration probably associated with electrolyte imbalances. Body weight was 20% lower at recovery than at launch. Ventricular fibrillation supervened 8 hr after recovery.

1. Introduction

In June 1969 a male *Macaca nemestrina* (pigtail macaque) was flown in earth orbit for 8.8 days by the US National Aeronautics and Space Administration, in collaboration with academic institutions. The flight was planned for as long as 30 days, but the central nervous and cardiovascular factors which combined to produce a progressive physiological deterioration appear to have general import in extrapolation to man. It is clear from manned flights by the USA and the USSR that man has not suffered as severely as did the monkey, Bonny, in Biosatellite 3. Nevertheless, the findings in Bonny appear unambiguous in showing effects of weightlessness on circadian rhythms, states of consciousness, central vascular functions and body fluid balance. All these effects have been described or presumed in man. The monkey flight provided the first direct measurements with implanted sensors of the physiological substrates of these changes. Findings in the monkey indicate that man may differ only in the severity of the changes, but not qualitatively. The benefits from such animal experiments are thus twofold. They serve to crystallize knowledge of complex physiological system interactions in the weightless state; and they point directly to areas of needed knowledge where man's ultimate weaknesses should be closely scrutinized. For predictive purposes, effective modeling of system limits must surely rest on a firm basis of corroborative observations in man and animals.

The Biosatellite 3 flight was formulated in 1964, initially as a study of effects of prolonged weightlessness on central nervous and cardiovascular functions. The central nervous studies proposed by the University of California at Los Angeles envisaged implantation of a *Macaca nemestrina* monkey with surface and deep brain electrodes for EEG recording, and with electro-oculographic (EOG) and cervical and trunk electromyographic (EMG) electrodes. In a joint experiment, the University of Southern California planned implantation of blood pressure sensors in central venous and central arterial structures. Data from these two experiments were to be correlated in evaluation of circadian rhythms, states of alertness and focused attention in task performance involving recent memory and skilled eye–hand coordination, and in fine analysis of sleep states. In 1965 additional experiments were added from other institution tutions. These included development of an in-flight urine analyzer by Jet Propulsion Laboratories for analysis of level of creatine, creatinine and calcium by the University of California at Berkeley. The latter group also contributed a whole-body metabolic experiment, which measured fluid distribution in extracellular and intracellular compartments pre- and post-flight. Texas Woman's University studied skeletal decalcification by X-ray wedge densitometry. Environmental radiation levels, specifically for high energy particles, were tested by the Donner Laboratory at University of California, Berkeley. Clearly, this was an unusually complex experiment, requiring close coordination from its inception, primarily by NASA personnel at Ames Research Center. After more than five years of preparation, Bonny was launched into orbit on 28 June 1969.

2. Résumé of Preparation of Flight Monkey

In its training and surgical implantation, the flight monkey followed a program identical with that in over 500 other monkeys [1]. All flight candidates were males weighing approximately 6 kg and physiologically just past puberty. On 1 December 1967 Bonny was transferred from the quarantine colony at NASA Ames Research Center to the UCLA Space Biology Laboratory, and underwent a six-month period of adaptation to collar restraint and daily handling. In May 1968 behavioral training in both flight tasks was initiated, and by 15 August 1968 this animal had received 67 one-hour training sessions [2]. Electrodes for EEG, EMG, EOG and brain temperature sensing were implanted 5 September 1968. From then until the flight, behavioral training and adaptation to couch restraint continued. On 11 June 1969 Bonny was selected as one of five flight candidates, all of which then went through the final countdown procedures, with surgery for a perineal urethrostomy on 15 June 1969 and implantation of vascular catheters and electrocardiograph leads on 20 June 1969.

It should be emphasized that, as with other flight candidates, the flight animal was accustomed to long periods of isolation and restraint each day in training capsules that closely simulated the flight vehicle. Training in the

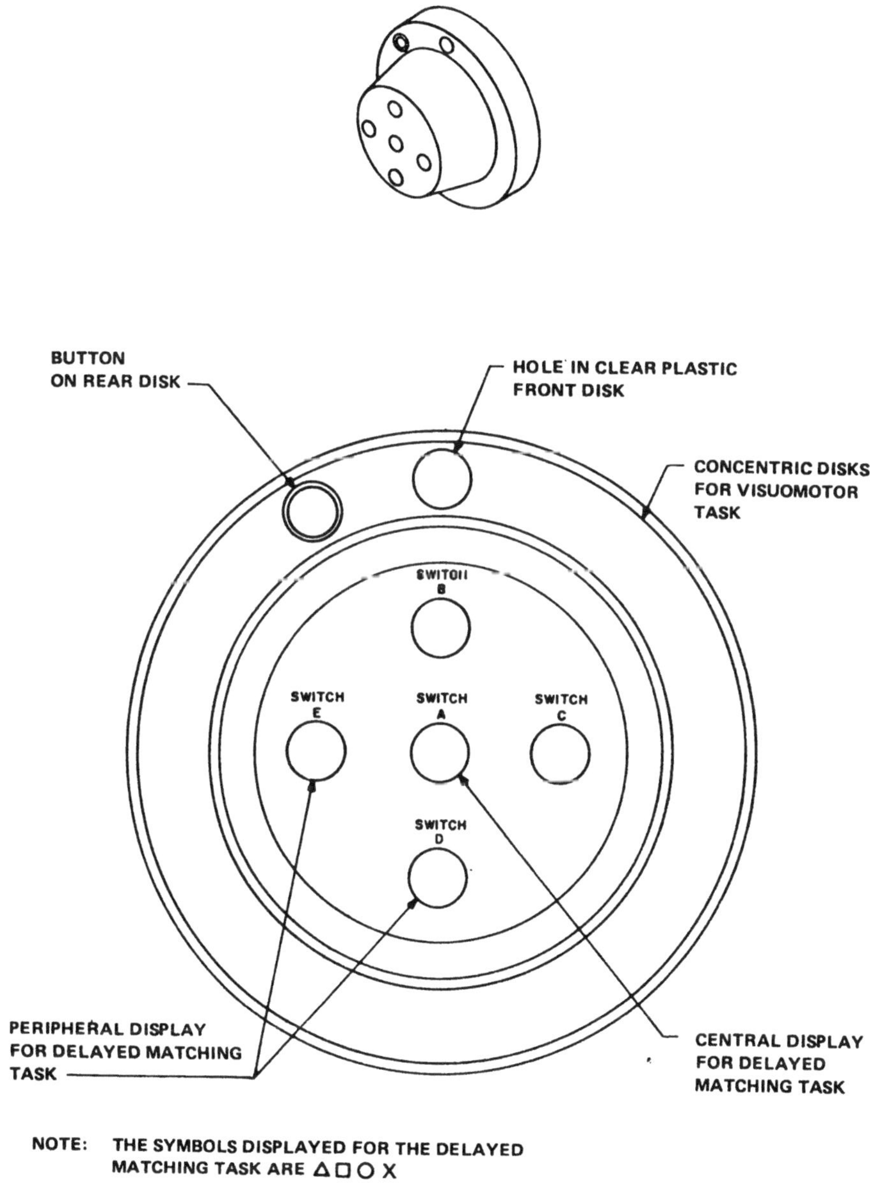

Fig. 1. Panel for behavioral tasks. This apparatus for displaying the two tasks was located on the primate's lap (from Adey and Hahn [1]).

behavioral tasks was also performed for several hours daily while in the simulator. Every effort was thus made adequately to adapt the animals for a long experiment [1]. Day/night lighting cycles also accurately reproduced spacecraft conditions for more than four months before flight, with 12 hours of "day" in 6 foot-candles of incandescent illumination, and 0.06 foot-candles of red light at "night". It was determined that these lighting schedules adequately entrained circadian rhythms in all ground tests. Behavioral tasks were presented by an apparatus consisting of five switches and two overlapping revolving disks (Fig. 1). The five switches were used in the delayed matching task (DM) and were transparent windows overlying neon tubes capable of displaying symbols. They were activated by touching the window. In a correct response, one of four different symbols was selected from a random matrix 18 seconds after it had been displayed separately. In the visuomotor task (VM), two disks co-rotating at slightly different speeds around 85 rev min^{-1} had brief periods of spatial coincidence of a window on one disk with a button on the other. The monkey touched the button through the window during the brief coincidence. Twenty trials of DM followed by 20 trials of VM were presented twice daily.

3. Physiological Responses to Effects of Launch

Physiological responses to the stresses of launch have been described in detail [3]. For 11 hr prior to launch, continuously telemetered data showed the animal to be alert throughout. He performed both tasks and ate briskly the allotment of 20 food pellets. Lift-off occurred at 0315.59 GMT on 29 June 1969. Ascent was accompanied by anticipated strong arousal reactions in EEG, EOG and EMG records, and by a rise in heart rate from 170 to a maximum of 234 beats/min. In the phase of powered flight from 175 to 203 sec, acceleration forces increased from 4 to 7 g and were associated with a horizontal nystagmus in the EOG. Computer analysis revealed strong coherence between the nystagmic eye movements and EEG records from visual and parietal cortex.

The animal weathered the stresses of launch with minimal discomfort. Heart and respiration rates increased at the onset of each new experience and decreased shortly thereafter. During the last 5 min of powered flight, the animal's physiological indicators had returned to an awake and normally alert state.

4. Flight Camera Data on Sleep/Wake Activity Patterns

A 16 mm movie camera was mounted over the animal's left shoulder in the spacecraft and in terrestrial simulations. Its field of view included a 24 hr clock and date indicator. Pictures were taken either at 20 min intervals or in a cine mode at 4 frames/sec. Each frame was analyzed for sleep/wake states,

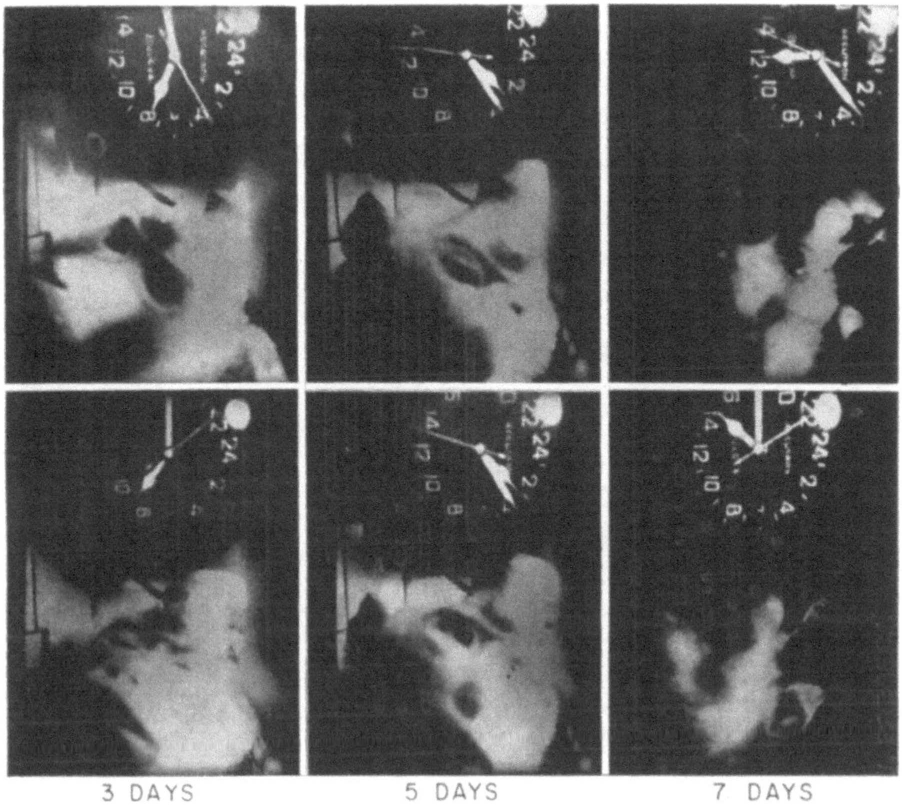

Fig. 2. The three pairs of photos from the flight camera showing the primate at different phases of the flight: Day 3, awake, then asleep; Day 5, during delayed matching (DM) task, showing head and eye movements; Day 9, cheek pouches filled and, 2 hr later, pouches empty (from Hoshizaki et al. [4]).

i.e. eyes closed – sleep; eyes open – wake; eyes not discernible – no data (Fig 2). Interpretations were supplemented by concurrent food and water consumption [4]. In control studies, with one exception, the sleep/wake activities were clearly synchronized to the imposed dark/light cycles. In the one exception the sporadic sleep/wake pattern was probably induced by problems in the animal's gaseous environment.

The camera data indicated that weightlessness does not affect the period length of the sleep/wake cycle. By combining telemetry data for food and water consumption with the sleep/wake records from the camera, we concluded that Bonny apparently dozed or slept most of the time and awoke during task periods, or whenever water was available. It appears that game-playing with accompanying food reward, the *ad libitum* feed period just before the night period began, the imposed schedule of available drinking water and the daily light intensity changes all played a role in synchronizing the animal's sleep/wake activity patterns to a 24 hr period. The "Zeitgeber" effect of these factors has been documented [5, 6, 7]. However, during 9 days of weightlessness, Bonny's 24 hr sleep/wake rhythm exhibited a phase angle difference of 2 hr from the imposed night/day modes (Fig. 3). A rapid shifting back and forth from sleeping to awake states was very obvious during daytime periods. From comparison with heart rate, brain and body temperature, and pCO_2 data,

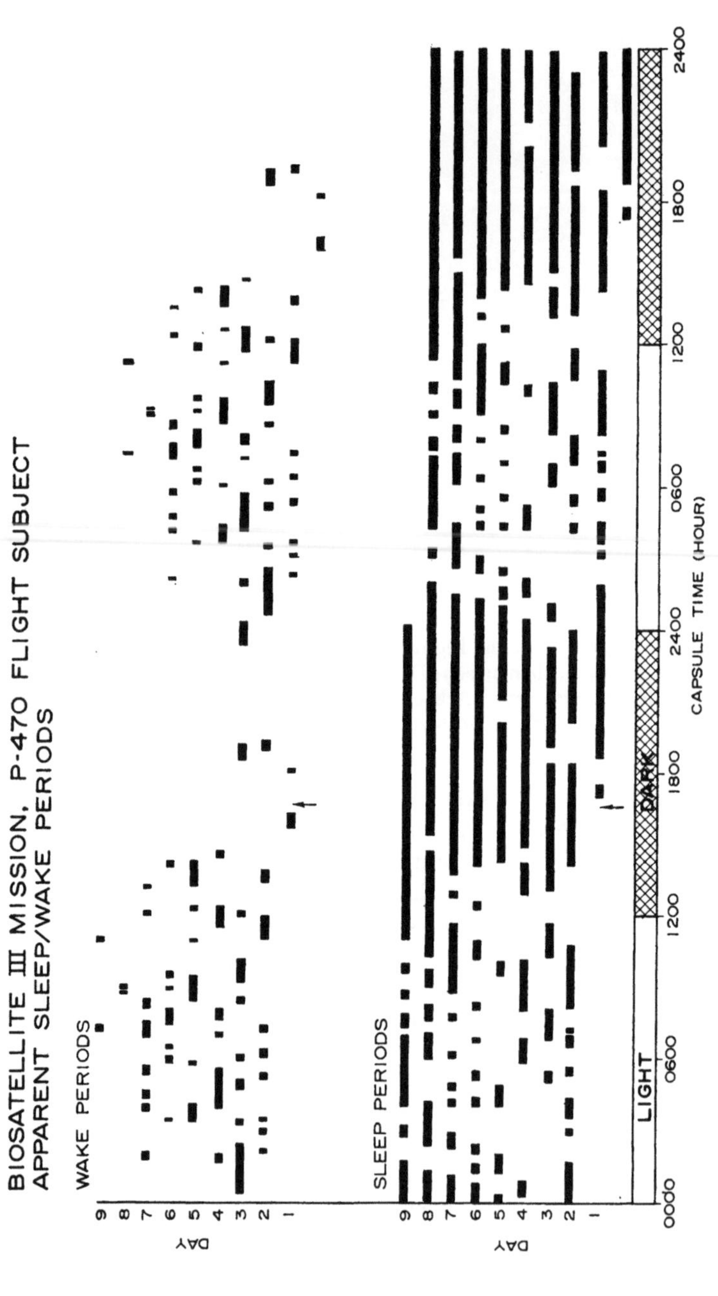

Fig. 3. Sleep/wake activities of the Biosatellite 3 monkey, derived from time-lapse pictures taken every 20 min at 0, 20 and 40 min after the hour. Each frame was considered to represent the animal's condition for that 20 min period. The sleep/wake states were scored as eyes closed — sleep; eyes open — wake; eyes not discernible — no data. To aid visual interpretation, the data are replotted on the right with a 1 day shift, to connect the data of consecutive days. The subject was generally awake between 0200 and 1400 capsule time, and shifted between sleep and wake states more than controls (Fig. 4). Wake periods fell dramatically on Days 8 and 9. Flight began on Day 1 at 1616 capsule time (arrow).

Capsule time of 0000 or 2400 is 0700 US Eastern Daylight Time, and is the beginning of the 12 hr "daylight" period.

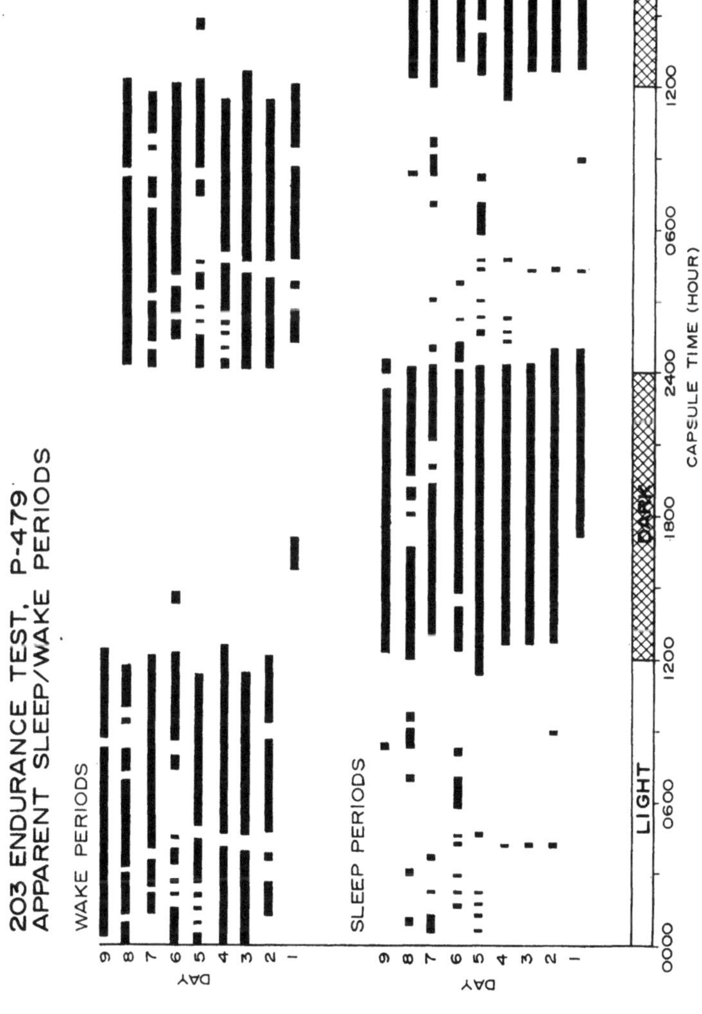

Fig. 4. Sleep/wake activity records from No. 203 Endurance Test (Monkey P479). The test was conducted in a capsule in an isolation chamber. The sleep/wake activity was closely synchronized to the imposed dark/light cycle (from Hoshizaki et al. [4]).

which exhibited a 26 hr rhythm [8] it is concluded that Bonny experienced an internal desynchronosis (Fig. 4).

5. Estimates of Circadian Rhythms in Bonny in Biosatellite 3

Daily or circadian rhythmicity in various bodily processes of mammals has been recognized for at least a century [9, 10, 11]. Rhythmicity of body activity, metabolism, excretion rates, thermoregulation and cardiovascular measures persist in terrestrial laboratory conditions where such environmental factors as light, temperature and humidity are locked to a 24 hr cycle. It is now believed that if these processes become arhythmic or desynchronized, a deterioration of the organism can result [12, 13], though this knowledge does not necessarily bring us significantly closer to an understanding of the cellular origins of biological clocks. Nevertheless, the possibility of desynchronosis is of great importance in planning the future of long manned spaceflight. Estimates of circadian rhythms have come to rest on sophisticated computer analysis for the finer indices. These include periodic regression [14], the Cosinor method [15], the Volterra integro-differential equation [16], and complex demodulation methods [17, 18]. Each method can give good results if at least 10 cycles of the variable are available. Here, the animal was weightless only 8.8 days and data during the last 30 hr showed rapid changes in the mean of each parameter which confused periodic analysis. The remaining 7.5 cycles of 24 hr rhythms are thus less than the desired number for good statistical analysis. Our estimation of period was in the most part by day averaging [8]. Data were interpolated to fixed 1.5 hr intervals to make an average for a 4-day period. Time displacement of two such tracings provides an indication of a non-24 hr period.

Environmental parameters, including magnetic field, total capsule pressure, relative humidity, capsule temperature and carbon dioxide tension, did not impress any unusual cycles on the animal. Brain and body temperature indicate a period greater than 25 hr. Similarly, the partial pressure of CO_2

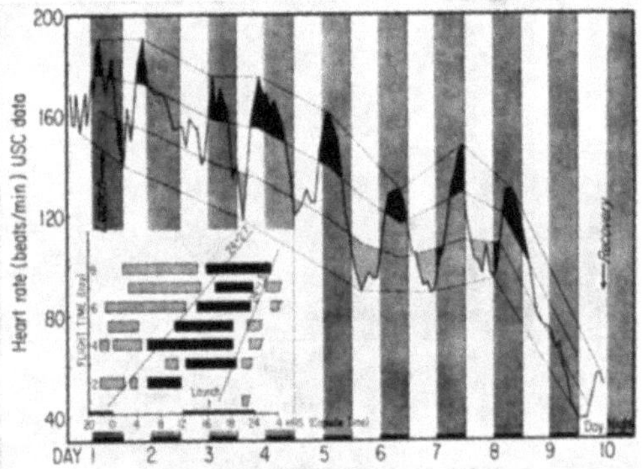

Fig. 5. Heart rate cycles on successive days of the flight. The inset at lower left shows the heart rate peaking later every day. The two linear regression lines indicate a daily period of greater than 24 hr (from Hahn et al. [8]).

and the heart rate, reflecting the animal's metabolism, showed a period greater than 25 hr. On the other hand, arterial blood pressure in the abdomen showed a rhythm of 24 hr (Fig. 5). The remaining four of the five subjects from which the flight candidate was selected were maintained in separate isolation chambers during the flight and showed no deviation from a 24 hr cycle in any recorded physiological parameter. Each of the four was subjected to the same rigorous countdown preparations and was "flown" to the profile of the flight animal with a 48 hr delay. Heart rate and brain and body temperature are impressive with a clear 24 hr cycle (Fig. 6).

The internal desynchronosis of the temperature, cardiac and respiratory cycles from the blood pressure on the one hand, and on the other, the external desynchronosis from the imposed 24 hr daily routine, may have been detrimental. The desynchronosis may have acted synergistically with the altered cardiovascular and fluid balance functions described below [19], to bring about the monkey's rapid deterioration.

The results noted during the flight have never been encountered in 125 animal-days of testing during simulated missions of 10–30 days with animals prepared in the same rigorous manner. This leads us to ponder the possible existence of a gravity-dependent mechanism in the control of circadian rhythms.

6. Sleep and Wake States estimated from Visual and Computer Analysis of EEG, EOG and EMG Data

Ten EEG, two EOG, and two EMG channels were among the 33 channels of physiological data monitored. EEG electrodes were placed bilaterally in the parietal and visual cortex, the hippocampus and the amygdala. Unilateral placements were made in the right centrum medianum and the left midbrain reticular formation. EOG leads were placed at the right and left outer canthi. EMG sensors were implanted in posterior cervical and lower scapular sites.

The sleep of the Biosatellite 3 primate in earth-orbital flight was abnormal in amount and bizarre in its distribution. It was characterized by unusually rapid transitions in state, notable brevity of state, and unusual transitions from one state to another (Fig. 7). The monkey never achieved its normal terrestrial cycle, and the remarkable fragmentation of consciousness required meticulous and virtually microscopic scoring of EEG records [20, 21]. Though Bonny retained essential features of his terrestrial sleep, with all sleep states detected at some time, scarely any aspect of his sleep was without atypical features. Rapid eye movement (REM) sleep, confirmed in space flight for the first time and typically associated with dreaming in man, was sharply reduced in amount, and associated with highly atypical spindling bursts (Fig. 8). Eye movements were also seen in sleep stages II and IV.

In Table 1, the reduction in REM and stage IV sleep is dramatic and obvious.

Table 2 compares the flight primate with baseline ground data from 20 consecutive nights' study of six other flight candidate monkeys. The reduction in REM sleep is even more striking than the table shows, since the 11% REM

a

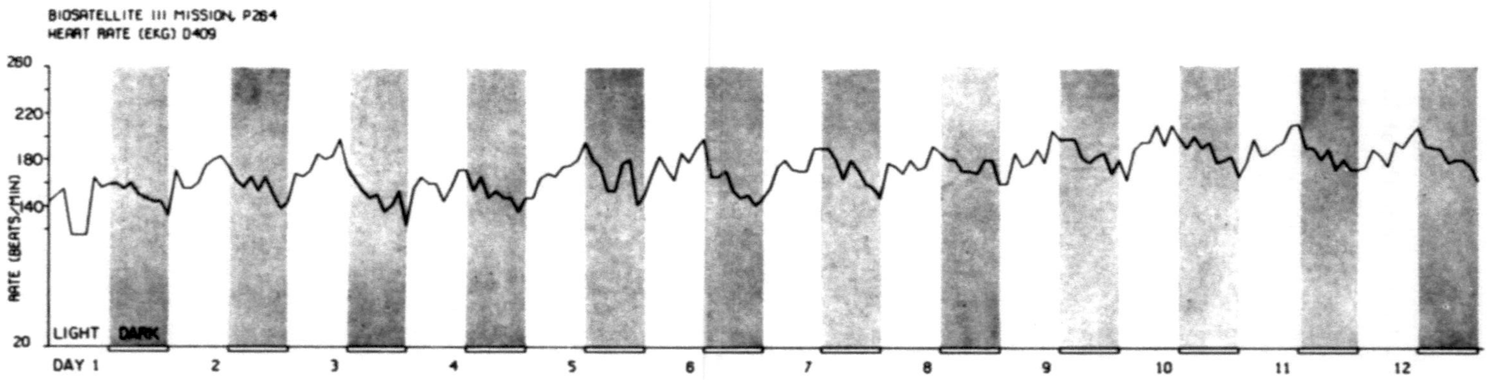
BIOSATELLITE III MISSION, P264
HEART RATE (EKG) D409
RATE (BEATS/MIN)
260
220
180
140
20
LIGHT
DARK
DAY 1
2
3
4
5
6
7
8
9
10
11
12

b

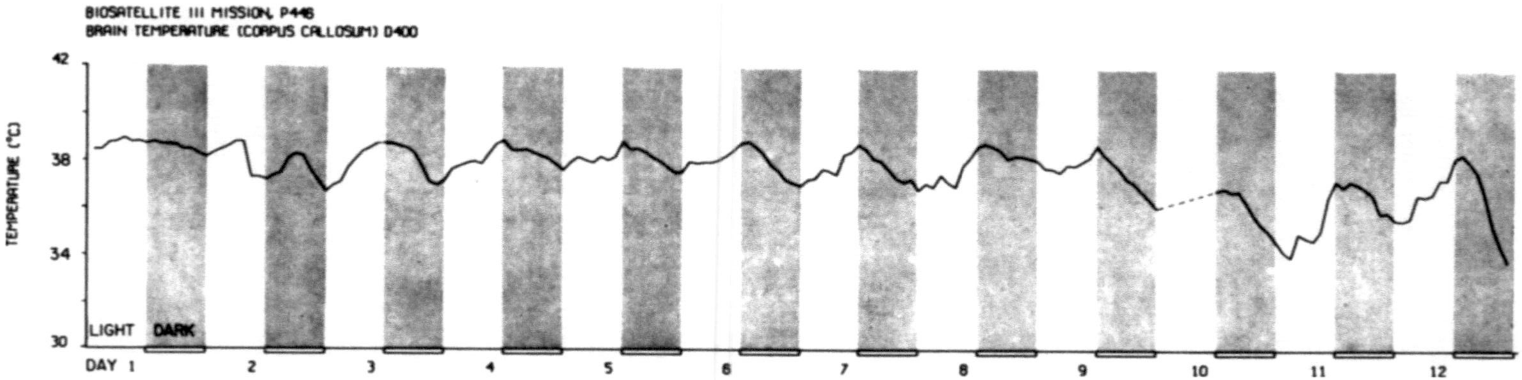
BIOSATELLITE III MISSION, P446
BRAIN TEMPERATURE (CORPUS CALLOSUM) D400
TEMPERATURE (°C)
42
38
34
30
LIGHT
DARK
DAY 1
2
3
4
5
6
7
8
9
10
11
12

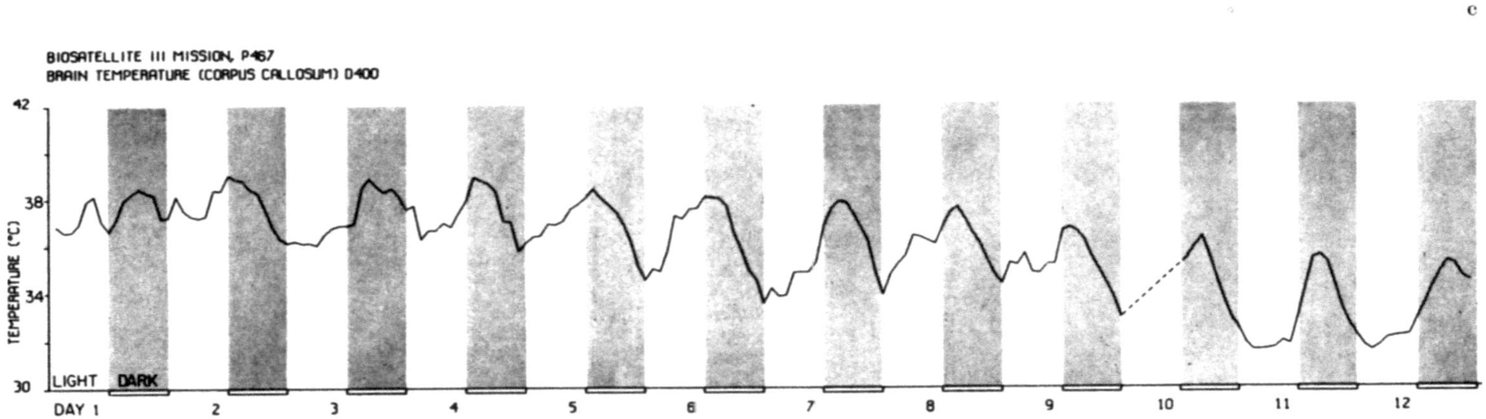

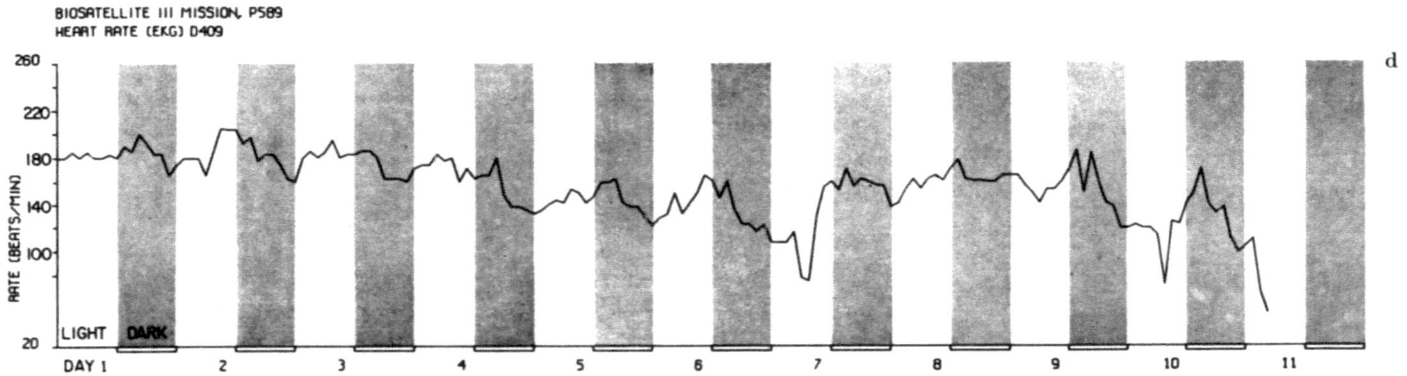

Fig. 6. The four remaining flight candidates were "logged" 48 hr behind the flight subject. The 24 hr rhythms are clearly identifiable (from Hahn et al. [8]).

Table 1

Flight primate sleep/wake staging

	%
Wake	45.6
Drowsy	20.1
Stage II	19.1
Stage III	7.6
Stage IV	3.2
REM	4.4

Table 2

Comparison of sleep staging between flight primate and base-line primates

	Base-line (%)	Flight (%)
Drowsy	14.6	36.9
Stage II	48.7	35.1
Stage III	12.0	14.0
Stage IV	13.6	5.9
REM	11.1	8.1

recorded in the animals on the ground is unusually low [22]. It is evident that the weightless state had a profound effect on the sleep and wake cycles of the animal. The chief characteristic, the remarkable fragmentation of consciousness, has been seen by us in only two circumstances: at night in the sleep of human quadriplegic patients with high cervical cord transections and

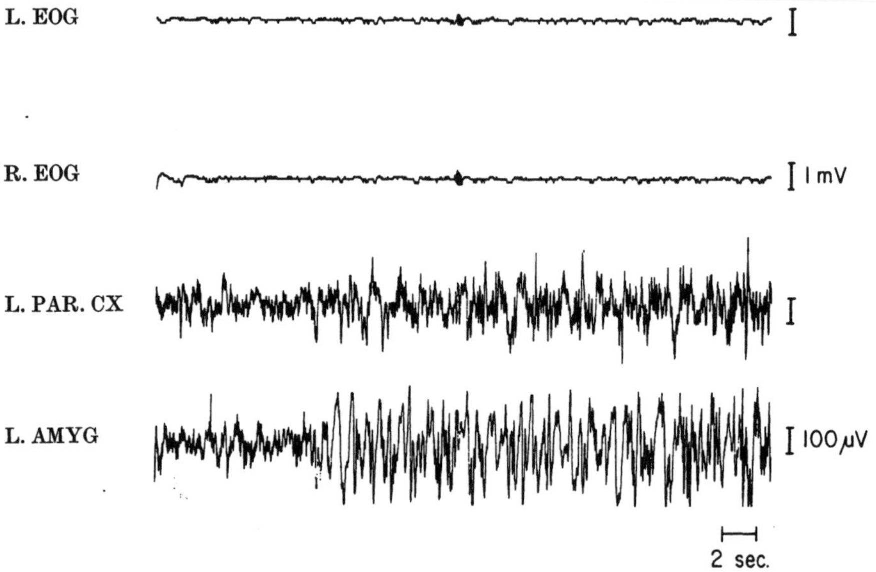

Fig. 7. Unusual transitions in EEG correlates of waking and sleeping, showing a waking/slow wave sleep transition in the left parietal cortex (L. PAR. CX) and left amygdala (L. AMYG). L and R electro-oculograms also shown (L. EOG and R. EOG). Orbit 67, Santiago, Chile: 3 July 1969, 1029 GMT (from Hahn et al. [8]).

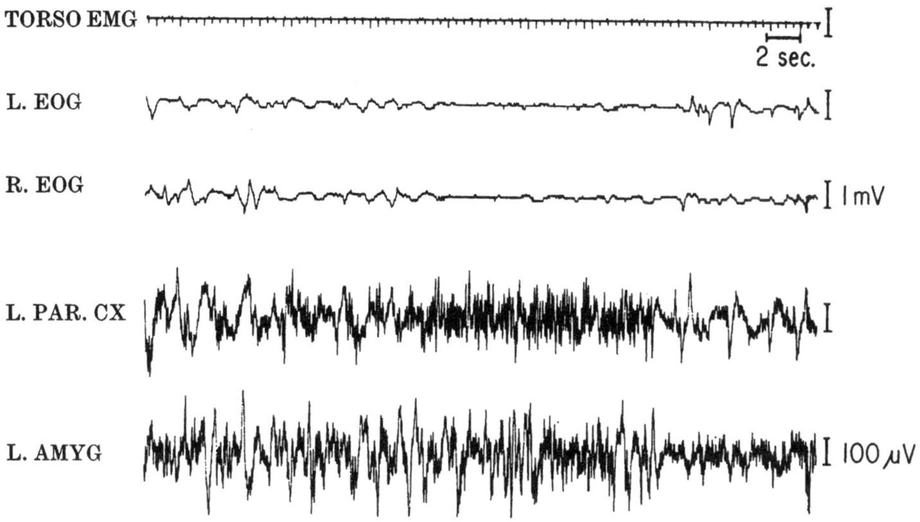

Fig. 8. Rapid eye movement (REM) sleep with highly atypical spindle burst in EEG record from left parietal cortex (L. PAR. CX). Orbit 37, Santiago, Chile: 1 July 1969, 1237 GMT.

during the day in the EEG of Astronaut Borman in the Gemini GT-7 space flight. On the supposition that the diminished somatosensory input in the high cervical cord transection might approximate the loss of such input in the weightless state, Adey et al. [23] studied sleep in these patients. They found rapid transitions in state and marked reductions of REM, stage III and stage IV sleep. Comparison between normal man, man with high cervical transection and the Biosatellite 3 monkey show that the sleep of the flight primate lies between that of normal man and man with high cervical cord transection, and approaches the latter. Rapid transitions between drowsy and wake states also characterized the Day 2 recording from Astronaut Borman during the Gemini GT-7 earth-orbital flight (Fig. 9).

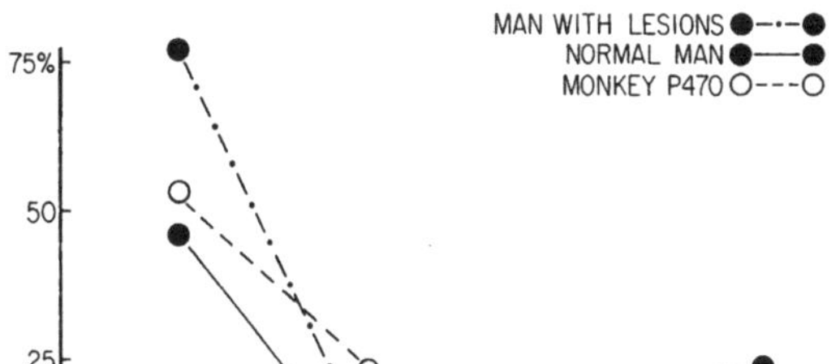

Fig. 9. Comparison of the amount of sleep spent in each of the sleep stages and the REM stage between the Biosatellite 3 monkey, normal man, and man with a high cervical transection (from Hanley and Adey [20]).

Gradual diminution and final cessation of eye movement note din the monkey may be important, since a similar trend was noted in Soviet cosmonauts in Vostok and Voskod spaceships [24]. Soviet investigators also noted that as the flights progressed, eye movements changed from frequent extreme deviations to rapid, bunched deviations of different amplitudes.

7. Observations on Cardiovascular Status of Bonny in Biosatellite 3

The basic premise of the vascular experiment [19] was that in weightlessness, as a consequence of reduction of the gravitational effect on the long columns of blood in the body, pooling of blood in the large vessels of the thorax would

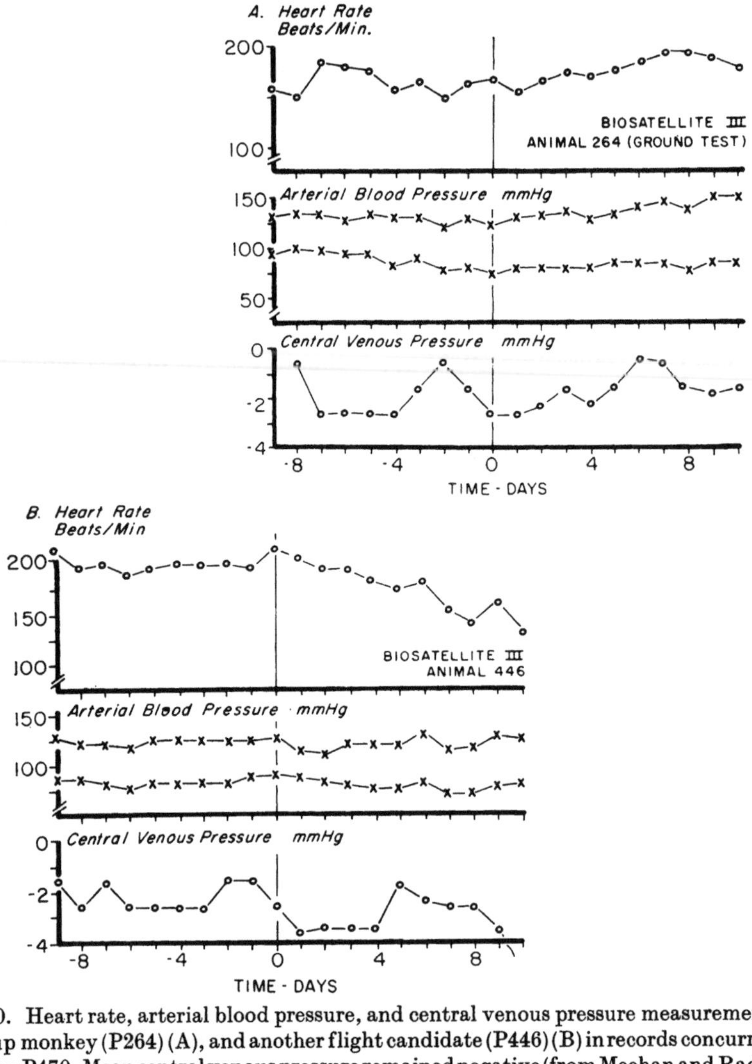

Fig. 10. Heart rate, arterial blood pressure, and central venous pressure measurements from back-up monkey (P264) (A), and another flight candidate (P446) (B) in records concurrent with flight by P470. Mean central venous pressure remained negative (from Meehan and Rader [19]).

occur; and because of this apparent high blood volume as indicated by stretch receptors in the large venous vessels, a compensatory mechanism would act to decrease the blood volume. This reflex mechanism has been shown to be operational in bed rest and water immersion experiments, and was predicted to operate also in reduced gravity fields. An indicator of this Henry–Gauer reflex is the pressure in the large veins near the heart. Catheters in these regions revealed for the first time an initial increase in pressure in weightlessness. One venous catheter was placed in the right atrium of the heart and the other at the entrance to the right atrium. Arterial pressures were measured in the abdominal aorta or common iliac arteries (Fig. 10). Initial orbital values of central venous pressure were increased approximately 2 mm Hg,

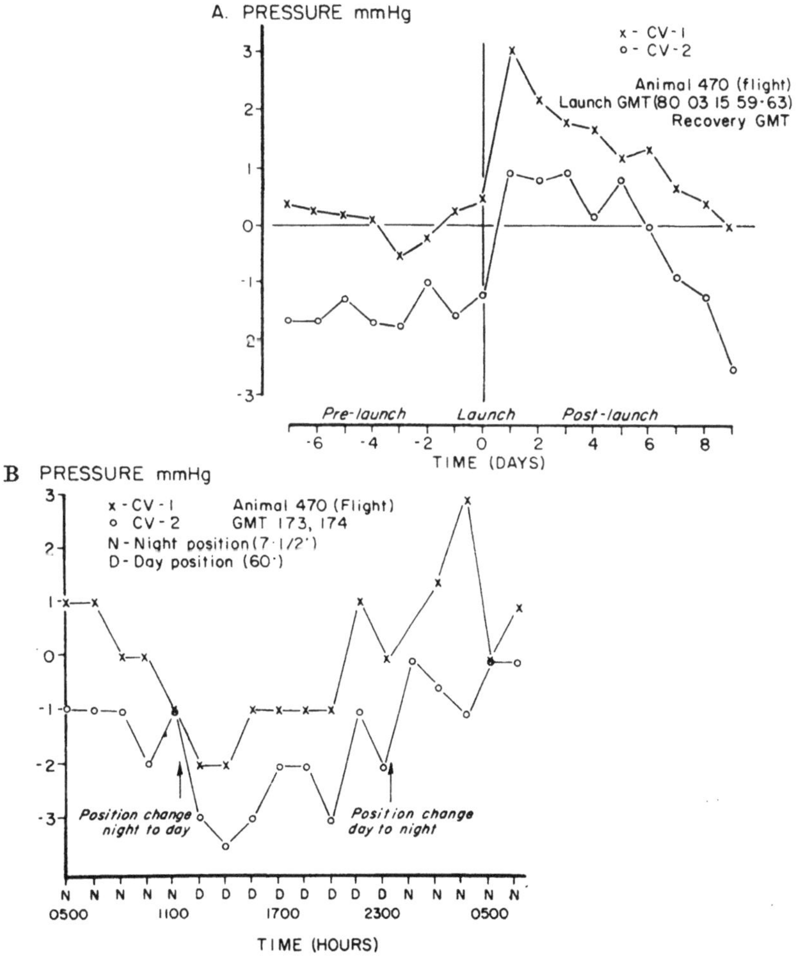

Fig. 11. A, mean central venous pressures in flight monkey pre-flight and during flight; B, changes in central venous pressure associated with diurnal shifts in body position of flight primate prior to launch (from Meehan and Rader [19]).

and this is of the same magnitude as changing from a sitting to a recumbent position (day-to-night shift in ground-based animals). The heart rate fell steadily throughout the flight. However, the arterial blood pressure was maintained at physiological levels into the eighth day. The average venous pressure fell steadily after the fifth day, but did not return to pre-flight values until the eighth day (Fig. 11).

Meehan and Rader [19] have evaluated fluid loss in the flight monkey. From a launch weight of 5.543 kg, there was a decline to a recovery weight of 4.430 kg. Estimating the water balance from intake, urine output, fecal loss and evaporative loss, they conclude that the negative water balance over the whole flight was approximately 1250 ml, and that the animal realized close to the maximum fluid deficit after 3–4 days of flight (Fig. 12). The relatively high urine volume in the face of a progressively increasing loss of body fluids strongly suggests that the animal encountered serious electrolyte problems. As shown by water immersion studies, the fluid loss will be hypotonic in the well-hydrated subject. However, when hydration is limited, the increased volume elimination is accomplished by an increase in osmolar clearance only, and excessive sodium loss can occur [25, 26]. This may be more severe in anxiety states. If the monkey were disturbed by his environment, irrevocable physiological derangements probably occurred early in flight, and thereafter, adaptation may not have been possible in a situation with such multifaceted stress.

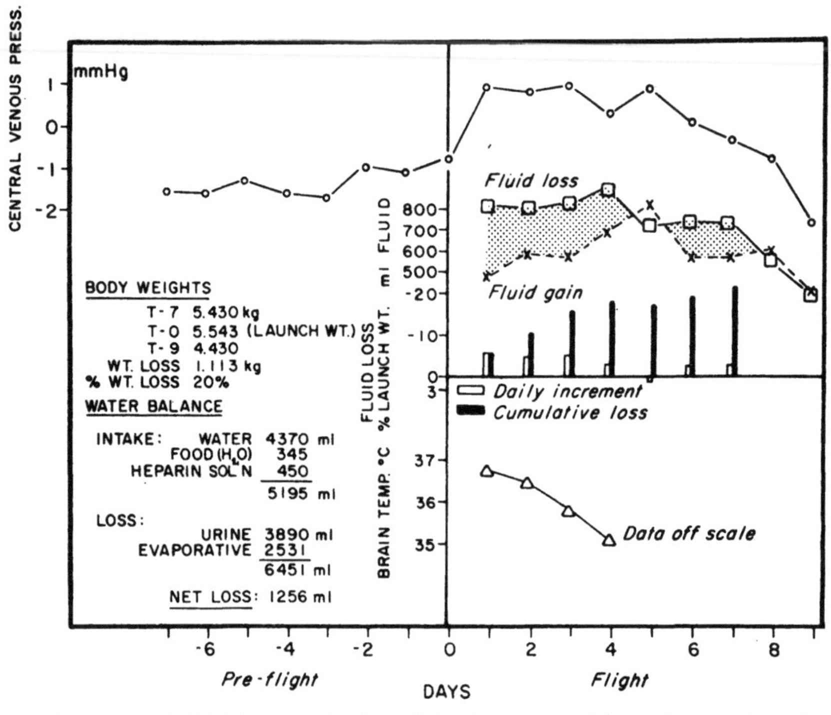

Fig. 12. Summary of fluid balance, body weight changes, and brain temperature during flight of Primate P470 (from Meehan and Rader [19]).

8. Bone Density Changes in the Flight Monkey

Mack [27] extended her observations of bone density changes in astronauts with detailed observations in Bonny, utilizing X-ray wedge densitometry measurements in 17 sites, including the calcaneum and talus in the ankle, the long bones of the upper and lower limbs, the phalanges of the hand, and lumbar and thoracic vertebrae. Measurements in the flight animal were compared with those in the back-up monkey. Both animals showed decreased densities over the same time span. In the control animal, these were attributable to immobilization. However, the changes were significantly greater in the flight animal in most locations, so that weightlessness exerts an added effect over simple immobilization.

9. Deorbit and Recovery Procedures

By the ninth day of flight, the heart rate had dropped to 60 beats/min and irregularities in rhythm occurred at intervals. The animal took adequate food and water on the ninth day, but the EEG resembled that of a hibernating animal [28]. By orbit 126, brain temperature had fallen below 35 °C. When the capsule was opened at Hickam Field, Hawaii, the primate was cold with scarcely perceptible respiration or heart beat. Resuscitative measures were immediately instituted, including intravenous dextran and cautious warming. After 3 hr, a regular heart action at 96 beats/min was established, and blood pressure, though low, was stable at 68/25, but remained quite sensitive to continued dextran infusion. Brain temperature had risen to 34.5 °C. His condition continued to improve and 7 hr after recovery, body temperature was 33.4 °C. He made coordinated stretching movements, his pupils were small and reacted to light directly and consensually, and EEG records resembled those in light sleep. Ventricular fibrillation supervened without warning 8 hr after recovery and did not respond to emergency measures. Approximately 90 ml urine was secreted in the post-recovery period.

10. Conclusions

The value of this animal experiment surely has its justification in the sensitive indices that it has provided, both in terms of subject susceptibility and the accuracy of the physiological data. It may be argued that since man has already endured the weightless state longer than this flight monkey, therefore animal flights are irrelevant. This view, while understandable, is specious and overlooks several important considerations. While it appears that *nemestrina* macaque is more susceptible to the hazards of the weightless environment, it is certainly an indicator of the possible hazards to man in long-term flight if suitable precautions are not taken, based on knowledge so gathered from general studies of mammalian systems in space.

There is clearly evidence that man to a lesser degree has experienced the changed physiology noted in the primate flight: cardiac embarrassment, vascomotor instability and increased fluid loss experienced in extravehicular activity in the Gemini GT-11 flight have been described in detail [29–32]; substantial weight loss, including some loss of muscle tissue, has been reported [33]; astronauts have complained of poor sleep, leading to procedural error which forced "real time" changes in flight plan: these sleep problems have required medication and the need for sleep EEG monitoring has been acknowledged [34]; frequent misperceptions of body position in space, subjective reports of illusory motion, including sensations reminiscent of the rotary chair, and the nystagmoid EOG with increased EEG slow waves in Cosmonaut Tereshkova, indicate vestibular disturbances in weightlessness [24, 35]; frank motion sickness has occurred on six occasions in the Apollo flights: three of these bouts required treatment and the problem has recently been conceded as significant [34].

The main findings in the Biosatellite 3 flight were:

(i) altered circadian rhythms, with external and internal desynchronosis

(ii) fragmented states of consciousness with rapid transitions

(iii) fragmented sleep states

(iv) altered vestibular functions accurately measured for the first time in US space flight

(v) modified central venous pressures and associated alterations in fluid balances.

All of these findings bear directly and positively on these problems in manned space flight. Man himself is a far less suitable subject for many of these measurements that directly assess his wellbeing.

Necessarily, such complex animal experiments must be carefully planned with a national purview. In this country, this review function has resided in part with the National Academy of Sciences and its Space Science Board. In a recent review entitled "Life Sciences in Space", a select committee of the Academy concluded, concerning the Biosatellite 3 flight, that if "considered solely as a contribution to fundamental physiological knowledge, the information to be obtained was not shown to be new and unpredictable, nor to require environmental conditions that were unique to space flight and that could not be adequately simulated on the ground." This opinion, offered in the name of the Academy even before the investigators had completed their flight data analyses, surely invites rejection as uninformed on the part of those who would advise government in the name of the scientific community. As scientists, we conclude that it were better if our peers and the bar of history were allowed to judge the merits of this extremely complex experiment which made unparalleled demands in dedication and unremitting endeavor. Surely the future demands a more constructive approach to investigation of life science problems in space than is manifest in the Academy report. Entry into space for measurements on man and animals remains difficult and expensive, so that experiment design must be in a frame of national and, hopefully, international cooperation. Observations on animals may be critically important in supplementing and broadening needed biomedical information about man, and in contributions to fundamental knowledge. Just as no man may enter the void of space by his own single endeavors, neither can any single biomedical experiment in space flight be justified except in the context of its contribution to a broad and carefully devised scheme of interrelated measurements, aimed collectively at a general understanding of man's capabilities and limitations in his extraterrestrial journeys.

References

[1] W. R. Adey and P. M. Hahn, Aerospace Med. **42**, 273 (1971).
[2] E. Campeau et al., Physiol. and Behav. **6**, 413 (1971).
[3] R. I. Tejada et al., Aerospace Med. **42**, 281 (1971).
[4] T. Hoshizaki et al., Aerospace Med. **42**, 288 (1971).
[5] V. G. Bruce, Cold Spring Harbor Symp., Quant. Biol., **25**, 29 (1960).
[6] J. L. Cloudsley-Thompson, Rhythmic Activity in Animal Physiology and Behavior, Academic Press, New York 1961.
[7] O. P. Stscherbakowa, in: Studeien über periodische Veränderungen physiologischer Funktionen des Organismus, Akad. Verlag., Berlin 1954 (p. 13).
[8] P. M. Hahn et al., Aerospace Med. **42**, 295 (1971).
[9] J. Davy, Phil. Trans. Roy. Soc. Lond. **135**, 319 (1845).
[10] C. Féré, C. R. Soc. Biol. (Paris) **40**, 740 (1888).
[11] W. Squire, Trans. Obstet. Soc. Lond. **9**, 129 (1867).
[12] F. Halberg, Réunion de Chronobiologie Appliquée à l'Hygiene de l'Environnement (Paris), June 1969.
[13] J. Harker, Cold Spring Harbor Symp., Quant. Biol. **25**, 279 (1960).
[14] C. I. Bliss, Conn. Agr. Exper. Sta. Bull., p. 615 (1958).
[15] F. Halberg et al., in: Cellular Aspects of Biorhythms, Symp. on Rhythmic Research, 81st Internat. Congr. Anat., Springer Verlag, Berlin 1967 (p. 20).
[16] C. M. Winget, D. H. Card and H. W. Hetherington, Aerospace Med. **39**, 350 (1968).
[17] D. O. Walter, Electroenceph. Clin. Neurophysiol., Suppl. **26**, 59 (1968).
[18] J. W. Tukey, Bull. Internat. Stat. Inst. **41**, 261 (1966).
[19] J. P. Meehan and R. D. Rader, Aerospace Med. **24**, 322 (1971).
[20] J. Hanley and W. R. Adey, Aerospace Med. **42**, 304 (1971).
[21] D. O. Walter et al., Aerospace Med. **42**, 314 (1971).
[22] M. L. Reite et al., Arch. Neurol. (Chicago) **12**, 133 (1965).
[23] W. R. Adey et al., Arch. Neurol. (Chicago) **19**, 377 (1968).
[24] O. Gazenko, in: Proc. 3rd Internat. Symp. Bioastronautics and Exploration of Space, San Antonio 1964 (p. 437).
[25] O. H. Gauer et al., Ann. Rev. Physiol. **32**, 547 (1970).
[26] J. A. Johnson et al., Amer. J. Physiol. **217**, 310 (1969).
[27] P. B. Mack, Aerospace Med., **42**, 828 (1971).
[28] F. Strumwasser, Ph. D. Thesis, Department of Zoology, University of California, Los Angeles, 1958 (p. 77).
[29] C. A. Berry, J. Amer. Med. Assoc. **201**, 232 (1967).
[30] NASA Gemini Summary Conference, Tech. Pap. NASA Symp. SP-138 (1967).
[31] NASA Gemini EVA Report, (NASA-MSC-G-R-67-2) (1967).
[32] G. F. Kelly et al., Aerospace Med. **39**, 611 (1968).
[33] C. A. Berry, Aerospace Med. **40**, 762 (1969).
[34] C. A. Berry, Aerospace Med. **41**, 500 (1970).
[35] C. W. White and C. A. Berry, Aerospace Med. **35**, 43 (1964).

Life Sciences and Space Research X — Akademie-Verlag, Berlin 1972

CALCIUM METABOLISM UNDER STRESS AND IN REPOSE

R. S. Goldsmith

Mayo Foundation, Rochester, Minn., USA

Derangement of calcium metabolism, although perhaps not as dramatic as that of the cardiovascular or vestibular systems, constitutes one of the major threats to the health of participants in exploration of space. On the basis of studies in immobilized subjects, the clinical disorders most likely to be encountered during prolonged space flight are primarily the consequence of an imbalance between bone formation and resorption (favoring the latter): (1) loss of skeletal mass, leading to osteoporosis; (ii) hypercalcemia; and (iii) hypercalciuria, with the attendant risk of nephrolithiasis. By itself, loss of skeletal mass would not be expected to pose an in-flight hazard, but hypercalcemia or nephrolithiasis could jeopardize lives or mission success.

Such data as are available from in-flight studies tend to support the use of immobilization as a terrestrial model for alterations in calcium metabolism during space flight. A variety of prophylactic measures have been attempted with this model in an effort to modify the observed disorders. Although there is some evidence that hypercalcemia and hypercalciuria can be reduced or prevented, negative calcium balance has not been completely reversed. Perhaps the most successful prophylactic measure utilized to date has been dietary supplementation of both calcium and inorganic phosphate.

With the wide variety of excellent study tools which are currently available for application to this field, significantly increased efforts are clearly required both to define the basic mechanism of immobilization-induced skeletal losses and to devise new prophylactic or therapeutic approaches.

1. Introduction

Through the ages, and indeed into the present century, it has been said that: "It is unnatural for man to fly. If God had wanted man to fly, He would have given him wings." I do not subscribe to this, but I am compelled to paraphrase it in this Space Age in a manner which accords with present physiological precepts: "It is unnatural for man to be weightless. If God had wanted man to be weightless, He would not have given him a spinal column." Perhaps this is an oversimplification, but not grossly so. The evidence is abundant that merely placing a normal man at bed rest, even allowing for such normal muscular activity as is possible in the prone or supine position, results all too rapidly in progressive dissolution of the normal skeleton. In Fig. 1, for example, is depicted the mean cumulative calcium balance of three normal subjects placed at bed rest for one month. Not only osteopenia, but several complications of the demineralizing* process, develop and endanger the very survival of the

* "Demineralization" or "demineralizing process" is used in this paper in the most general sense to mean resorption of bone. It is not intended to imply halisteresis.

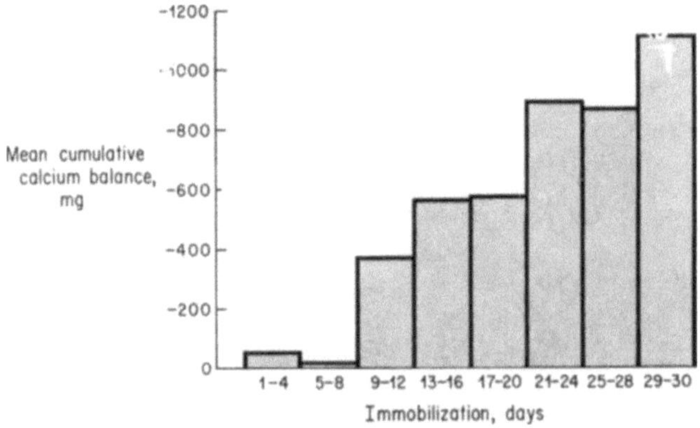

Fig. 1. Effect of bed rest on cumulative calcium balance. Height of bar represents mean values in three normal subjects.

organism. It is not possible to state with assurance, of course, that weightlessness can be equated with bed rest or immobilization, since (with one exception) data on calcium and phosphorus metabolism during prolonged space flight have not been obtained to date. From data on cardiovascular and vestibular function during such flight, however, there is a remarkable similarity to, and even exaggeration of, the effects of bed rest, and one can only speculate that the same may be true for bone.

2. Skeletal Function and Structure

Let us look at the need for a skeleton and the physical stresses which affect it. The primary function of the skeleton in man, aside from its being the principal site of hematopoiesis, is to provide a rigid framework for muscle actions. Fulcrums, levers, ball-and-socket and mortise joints, and many other well-known physical principles have been beautifully engineered by the extraordinary process of evolution. An excellent example of this engineering skill, and particularly germane to the subject of weight-bearing is the trabecular structure of the vertebrae. The principal trabeculae are oriented vertically, while thinner, less well-developed trabeculae course horizontally at right angles to the principal trabeculae. From a structural point of view, one can envision this as nearly ideal to resist compression without buckling [1–3]. Alternatively, the vertical trabeculae may be viewed primarily as "pressure" trabeculae and the horizontal ones as "tension" trabeculae [1]. In other words, during vertical compression (as with weight-bearing), there is a tendency for lateral expansion which is resisted by the "tension" trabeculae. With release of compression, elastic recoil produces an inversion of these "pressure" and "tension" relationships. Similar dependence of trabecular architecture upon functional need has been demonstrated for many bones, notably those of the lower extremity. It is all too clear from these various studies that architecture as we know it has evolved in response to the stresses of weight-bearing.

Of great significance to space travelers is the normal remodelling of the skeleton, which is constantly in progress. It has been demonstrated that bone is always replacing itself, with new bone being laid down along lines of stress. As an example, following initial healing of a misaligned fracture of a long bone, new lamellar bone is progressively added along the concave side of the fracture site while bone is being removed along the convex side. This rearrangement continues until a structure approximating the original bone shape is reconstructed. That this is due to stress rather than to any kind of memory by osteoblasts and osteoclasts is attested by the fact that this remodelling occurs only if the limb is being used in a normal fashion.

In 1892 Wolff [4] formulated a statement relating bone architecture to function, which is directly germane to our discussion. As amplified by Bassett just six years ago [5], Wolff's law states: "The form of the bone being given, the bone elements place or displace themselves in the *direction* of the functional pressure and increase or decrease their mass to reflect the *amount* of functional pressure". The elements of the skeleton are normally oriented so as best to resist compressive and tensile stresses, which are the forces ordinarily encountered on land. But it is essential that these elements be able both to reorient themselves and to change their mass to adapt the organism appropriately for altered environmental situations. We shall not examine macroscopic structure *per se*, nor all of the factors affecting response to deformational stress (Table 1). Rather, we shall confine the discussion primarily to long-term remodelling factors, involving cellular responses.

Table 1

Factors affecting response to deformational stress

Hardness
Visco-elasticity
Fatigue
Remodelling
  Osteoblasts
  Osteoclasts
  Osteocytes

It has been demonstrated that bone, a multicrystalline material, possesses piezoelectric properties [5–10]. In engineering parlance, therefore, bone behaves like a transducer, in that it converts energy from mechanical input into electrical signals (output). The precise locus of this transduction is not known, although there is some evidence that it may arise within the crystal lattice, perhaps at the collagen–hydroxyapatite crystal interface, which of course functions like a membrane [5, 10]. Of significant importance, electrical output from non-living strips of bone has been shown to be a nearly linear function of deformational stress, resembling the typical stress–strain relationship [5]. Since compressive and tensile forces are the major mechanical deformational stresses acting upon bone, and since these forces are opposite in direction to each other, one would anticipate generation of currents of opposite charge by these forces. If such were found to be the case, one would then have, in effect, a perfect example of autoregulation within the organ system itself. It has been demonstrated that both the amplitude and duration of electrical pulses generated within bone vary with the rate, magnitude, and duration of deformational stress, as well as with orientation of blood vessels, osteones, and lamellae, and with the degree of mineralization and hydration [5]. Furthermore, stress resulting in deformation or bending of a bone specimen has been found to produce negative polarity on the concave surface (compression) and positive polarity on the convex surface (tension). From Wolff's law and from many clinical and experimental studies, it is known that remodelling of bone occurs to increase the amount of bone on concave surfaces and to decrease that on convex surfaces. Thus, one might guess that electronegativity leads to increased osteoblastic activity and osteogenesis while electropositivity leads to increased osteoclastic activity and osteolysis. Without going into further detail about studies designed to answer this question, we can state that data have been accumulated which are consistent with this concept, but do not prove it conclusively [5]. The limitations may be methodologic.

Two qualifications of this discussion of piezoelectricity in bone must be mentioned. Firstly, although collagen has been demonstrated to possess piezoelectrical properties, no such properties have been demonstrated for hydroxyapatite. Nonetheless, there is substantial reason to believe that at least a portion of the piezoelectric effect may originate at the collagen–hydroxyapatite interface, and perhaps even within the hydroxyapatite crystal as well [6]. Secondly, some investigators have questioned whether bone should be thought of as a semiconductor and have suggested rather that it may be a "semi-insulator" [9]. This is a fine point which we cannot discuss in this presentation, but suffice it to say that bone possesses electrical properties which may have profound effects on function, particularly as related to remodelling. Thus, if one were to restate Wolff's law in modern terms, it might take the form shown in Table 2. Compressive stress is converted to electronegativity, which in turn induces osteogenesis, and the obverse is true for tensile stress.

Table 2

A modern version of Wolff's law

Stress/strain	Piezoelectric signal	Cellular activity
Compression	Negative	Osteoblastic
Tension	Positive	Osteoclastic

3. Complications of Immobilization

The complications of immobilization are skeletal demineralization, hypercalcemia, hypercalciuria and nephrolithiasis; these are considered in turn.

3.1. Demineralization

The importance of Wolff's concept to immobilization or its probable counterpart, weightlessness, is obvious. The normal mechanical stresses which are probably responsible for the majority of remodelling in living bone are: gravity (weight), muscle tone and activity, impact, and cardiovascular dynamics. It is evident that, with the possible exception of cardiovascular dynamics, compressive deformational stresses are largely obliterated by either bed rest or weightlessness. Perhaps even more important is the fact that absence of compressive stress is interpreted as a relative tensile stress, leading to less electronegativity or more electropositivity. In effect, then, lack of compressive stress of the sort listed at the beginning of this Section favors the resorption of bone by increased osteoclastic activity. Indeed, with no compensatory osteogenesis, it is difficult to imagine that there would necessarily be an endpoint other than complete resorption of the skeleton. In fact, of course, this is not likely to be the case. In some of the reported studies of immobilization [11–14], there appeared to be a decreasing rate of loss of calcium after the initial period of progressively increasing losses, suggesting the development of a new steady state. A mathematical model for predicting such behavior has been suggested by Stubbs [15]. He has pointed out that the proper interpretation of stress/strain relationship to bone remodelling entails a change in the ratio of osteoblastic to osteoclastic activity, which in turn alters the resistive architecture of the bone to decrease the strain by a given stress. When the strain has diminished as a result of the architectural change, the piezoelectric response has proportionately diminished, tending to bring the altered osteoblastic/osteoclastic ratio back toward control. Thus, the system may be considered convergent, in that the strain and the cellular response to it are always tending to decrease progressively toward a new set-point or equilibrium.

Vose and Keele have studied the roentgenographic density of the *os calcis*, using a sophisticated scanning technique [16], in a large number of mentally retarded patients who exhibited varying degrees of bedfastness and physical activity. Although it is not certain that the density of the heel is necessarily a reflection or predictable function of total skeletal, or even spinal, density their data clearly indicate that density was inversely related to the degree of hypokinesia, even in completely bedfast patients. Since impact on the heel as a result of ambulation could not be considered as a contributory factor in the fully bedfast patients, there is thus reasonable evidence that muscular activity is a major contributor to maintenance of normal skeletal architecture.

A recent observation of great potential interest was that reported at the European Bone Conference just a few months ago on the effect of a static magnetic field on collagen development *in vitro* [17]. The investigators presented preliminary data that there was a large increase in the amount of collagen produced by cultured fibroblasts when the culture dishes were placed in a static magnetic field. Whether these data pertain *in vivo* or have relevance to the potential problems of flight beyond normal terrestrial magnetic fields is totally unknown at this time.

There appears to be one major exception to the normal remodelling process, and that is of utmost significance to space flight: if a trabecula of bone is completely resorbed, there exists no template for a new trabecula to replace it. This is demonstrated most strikingly by the osteoporosis of the spine which occurs as a result of Cushing's syndrome [18]. When the Cushing's syndrome is completely cured by surgical removal of an adrenocortical adenoma, a peculiar kind of healing of the vertebrae occurs. The relatively unaffected cortex of the vertebral bodies increases in thickness and density, but the spongy or trabecular bone of the vertebrae remains as before, producing the so-called picture-frame appearance. Thus, loss of trabecular bone may proceed to such an extent that *healing is not possible*, even though the etiology has been completely removed. As a result, it is of paramount importance to define the degree of demineralization which may be caused by the weightless state and to attempt by every means possible, whether physical or pharmacologic, to prevent such demineralization.

3.2. Hypercalcemia

Probably of lesser significance in terms of long-term effects, but possibly of substantial in-flight danger, is the potential development of hypercalcemia. In virtually every study of immobilization, an incidence of hypercalcemia of about 10% has been observed. In general, these elevations of serum calcium have been small and have not been considered dangerous to the immobilized patient. One reservation, however: immobilized patients are not called upon to make razor-sharp decisions or precisely timed course corrections. The serious problem of even mild hypercalcemia is that it may produce apathy and inability to concentrate, nausea and vomiting, and less commonly stupor and reflex changes. More severe hypercalcemia, of course, can cause coma, convulsions, and even death. Any degree of hypercalcemia leads to changes in impulse conduction through the heart, primarily an increase in the time from ventricular contraction to repolarization. As a result, hypercalcemia could, at the least, lead to a reduction in crew efficiency and, at the worst, to serious morbidity or even mortality.

3.3. Hypercalciuria and Nephrolithiasis

Hypercalciuria or increased urinary excretion of calcium, is a nearly universal accompaniment of immobilization (Fig. 2) and was observed in one of the two astronauts studied during the flight of Gemini 7. In itself, hypercalciuria may not be hazardous, although there is a suggestion in the literature that it may lead to a reduction in renal function [19]. Undoubtedly a serious consequence of hypercalciuria, however, is its potential for promoting nephrolithiasis, or renal stone. Although there are many factors which contribute to renal stone disease, there is no doubt that the concentration of urinary calcium plays a major role. With the dehydration and decreased urinary flow which may

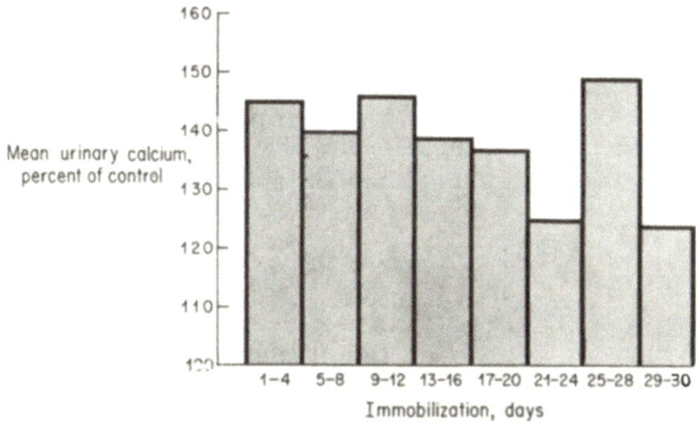

Fig. 2. Effect of bed rest on urinary excretion of calcium. Height of bar represents mean values (as percentage of control values) in three normal subjects.

accompany space flight, concentration of calcium may be increased even further. In general, renal stones develop slowly, but a multiplicity of factors, some known and others unknown, may greatly accelerate their production. Patients with certain kinds of renal tubular disorders involving inability to acidify the urine and accompanied by hypercalciuria, for example, may produce stones at the rate of several per week. Renal stone may not conjure up any serious problems to the uninitiated, but one has only to experience the pain or witness it in a patient during passage of seemingly minute stones to realize that this is probably the most painful of all human experiences. Since the incidence of renal stone approximates 25% in patients immobilized by paraplegia, we are dealing with a potential threat of substantial probability of occurrence. Perhaps of more long-term consequence than the acute disability of stone passage are the twin sequelae of infection and reduced renal function, which are frequently encountered in the presence of renal stone.

3.4. Muscle Wasting

What of the possibility that muscular strength and coordination may be impaired by prolonged weightlessness? This subject is mentioned because of its relation to the skeleton, but only briefly since it is peripheral to our discussion of calcium metabolism. We are all aware of the gross lower extremity weakness, and even wasting, which may occur with prolonged immobilization, as witness the faltering first steps of the patient recovering from prolonged illness in bed. It must be self-evident that, just as the skeleton wastes from lack of gravity stress, so does the musculature involved in normal locomotion. Such effects have been demonstrated in most studies of immobilization, and suggestion of a similar effect may be inferred from some of the data derived from the studies in Gemini 7 astronauts [20]. I have used the term "inferred" because of the indirect nature of the evidence. Both subjects manifested an increase in urinary excretion of inorganic phosphate which could not be accounted for by resorption of hydroxyapatite from the skeleton. Because of the difficulties encountered during execution of the studies, it was not possible to calculate "theoretical" phosphate balance [21] and thus to estimate loss of lean body mass or muscle. The increases were of sufficient magnitude, however, coupled with the comment of one of the astronauts on the "flabbiness" of his leg muscles, to suggest the possibility of significant loss of muscle mass.

4. Study Techniques

4.1. Serum Calcium

What are the techniques available to us for study of the effects of weightlessness? Fortunately, there is a wide variety of such techniques for examining virtually all aspects of calcium and bone metabolism, and the factors which influence them. The difficulty, of course, is that many of the methods would be difficult to apply within the constraints imposed by space flight, and certain answers will have to be determined either by analogy to immobilization or *post hoc*, after flight; some of the methods have been attempted on past flights and others are planned for future flights, such as in the orbiting laboratory setting. Taking the potential problems singly, it will become clear that many of the answers can be resolved. Firstly, the possibility of hypercalcemia is difficult to examine directly during flight because of the requirement for withdrawal and processing of blood samples. An alternative approach, however, which is probably in use already, is the evaluation of certain portions of the electrocardiogram. The plasma concentration of calcium has a relatively specific effect on the duration of the ST segment, and an excellent negative regression can be demonstrated [22]. Because the ST segment is so short, the usual measurement made is the $Q\text{–}T_c$ interval, i.e. from the beginning of Q wave to the end of the T, corrected for rate. The problem with this measurement is that it includes the T wave, which is subject to a wide variety of influences. Bronsky [22] has suggested, therefore, that the $Q\text{–}{}_0T_c$ interval be utilized instead. This interval terminates with the onset of the T wave and is less easily affected by other factors. My own experience in a large number of hypercalcemic patients corroborates this observation, and we have often been able to predict plasma calcium concentration quite accurately from this index. Since in-flight electrocardiographic monitoring is routine, this assessment obviously presents no serious obstacle.

4.2. Urinary Calcium and Stone

More difficult, however, is the evaluation of hypercalciuria and potential for stone disease, which cannot be determined indirectly. A previous attempt to collect urine specimens in-flight by a rather elaborate system, failed because of improper functioning of the system [20]. Subsequent evaluation of urinary calcium had to be based on the assumption (probably reasonable) that glomerular filtration rate and urinary excretion of creatinine had not changed. The problems of such an assumption are that dietary intake of pre-formed creatinine was not adequately controlled, and there was probably an increase in muscle breakdown (as evidenced by increased phosphate excretion). These two factors would tend to affect urinary excretion of creatinine in opposite directions and may very well have cancelled out. In any case, more accurate collection of specimens is necessary and should be readily solved within currently available engineering procedure. Of far greater import is the question of stone potential, which is a variable under active investigation, but not yet resolved. Many factors are known to influence the propensity to develop urinary calculi, including urinary excretion of calcium, phosphate, magnesium, pyrophosphate, and an inhibitor substance of as yet unidentified composition [23, 24]. All of these factors interact, but the "inhibitor substance" seems to play a central, and perhaps dominant, role. Fortunately, the presence or absence of this substance can be determined readily by a simple *in vitro* system which depends upon whether or not urine permits or prevents calcification of rachitic cartilage or reconstituted collagen [23, 24]. It would be desirable, therefore, to include such an assessment in the examination of all candidates for prolonged space flight. Perhaps even more important is the fact that a simple dietary maneuver (see below) can simultaneously decrease urinary calcium, increase urinary pyrophosphate, and cause the inhibitor to appear — all three of which "stabilize" the urine.

4.3. Demineralization

More difficult to evaluate, yet probably the most important for the long term, is the effect of space flight on the skeleton. A variety of techniques have been employed to make this assessment, with less than satisfactory results to date for the purpose of space flight. The determination of calcium balance, i.e. the difference between calcium intake and excretion, is a laborious tool which has served the calcium community well during long years of experience. Since by far most of the body calcium is in the skeleton, calcium balance reflects almost exclusively skeletal balance, the sole exception being the situation in which calcium is being removed from bone and deposited in tissues which are not normally calcified. Studies during enforced bed rest in normal subjects have consistently demonstrated negative calcium balance, which persists for long periods. The single attempt to examine calcium balance during orbital space flight (Gemini 7) was unsuccessful for a variety of reasons [20]. It did demonstrate that excretion of calcium increased during flight; the technique of calcium balance, however, requires too great attention to detail and is too cumbersome for practical utilization during prolonged flight, although it is planned to perform such studies in the orbiting laboratory program. Such data will be invaluable at this stage, primarily because of our need to know if enforced bed rest is, in fact, a suitable model for weightlessness. One addition to the planned protocol should be considered in my opinion, and that is the utilization of a quantitative recovery marker, such as chromic oxide or polyethylene glycol. Such an addition would permit correction for possible losses of fecal specimens without resultant compromise of data.

One fascinating possibility for assessing change in total body calcium without the need for balance studies is the technique of total body *in vivo* neutron activation, which has recently been employed for some studies of osteoporosis [25]. The technique appears to be a valid one, but has the disadvantage that the required instrumentation limits measurements to pre- and post-flight examination. Thus, it would not be possible to determine the time-course of change during flight, but it *would* be possible to examine the time-course of recovery, even over an interval of months, a feat which is simply not practical for the balance technique. In addition, it would be desirable to assess rates of calcium accretion and resorption. Although the technique of radio-calcium kinetics, as normally performed, requires serial blood sampling, it is possible to make the appropriate calculations from urine specimens, providing no samples are lost and time of voiding is accurately recorded.

Although bone biopsy would provide yet another parameter of skeletal response, this is obviously impractical. Alternative solutions to more direct assessment of bone *per se* have become available in the form of a variety of non-invasive techniques for evaluating bone mineralization. It is evident that standard X-ray techniques are totally unsatisfactory for assessing skeletal mineralization, since we are interested in quite small degrees of change. During recent years, several techniques have been described which are claimed to have reproducibility several-fold better than standard X-rays (2–3% compared with 10% for X-rays). The major limitation of all such techniques so far has been their dependence upon use of the appendicular, rather than the axial, skeleton. Since the spine is probably the portion of the skeleton which is of greatest concern, at least from the point of view of ultimate recovery or residual disability, it is essential to determine the degree to which the techniques in use reflect the axial skeleton. Such information has been difficult to obtain because of the lack of a satisfactory reference standard for evaluating the spine.

Each technique has its advantages and disadvantages, and some have had reasonably extensive use in various clinical situations, but none has yet satisfied completely the criterion above. Mack and her associates have examined astronauts before and after three of the flight and have found definite evidence of demineralization of the calcaneus after the flights [26]. The degree of demineralization did not correlate with the duration of flight, but this was interpreted to be due to different amounts of physical exercise and diet. It is unknown at this time whether such conclusions are warranted or will be confirmed by subsequent studies. As with total body neutron activation, assessment of bone mineralization by these various techniques is limited to pre- and post-flight times. Perhaps even more serious a limitation on the methods as currently employed is that they tell us nothing about the actual architecture of the bone, but only the amount of calcium or degree of mineralization. It would be at least as important to know the macro- and microarchitectural alterations which occur. Not only would this be useful for evaluating the effect of space flight itself, but might provide clues to the type of therapeutic or prophylactic maneuver(s) which might prove useful as well as to potential reversibility.

Deitrick et al., in now classical studies [12], demonstrated that bed rest led to large losses of calcium from the skeleton, both by the urinary and fecal routes. Findings were similar in both normal subjects and patients with paralysis due to poliomyelitis. In the normal subjects, however, the extreme negative calcium balance could be mitigated to some extent by the muscular activity produced by a rocking bed, i.e. by the involuntary alternate tensing and relaxation due to tilting. In addition, normal subjects promptly developed positive calcium balance upon resumption of ambulation. The degree to which skeletal repair occurred, however, could not be assessed in those studies, except by inference from calcium retention.

Heaney [13] has reported on the calcium dynamics which occur during disuse osteoporosis. He found that, during the acute or active phase of disuse, there was a large increase in turnover of skeletal calcium. Both calcium accretion and resorption were increased, but resorption rate was more than 50% higher than accretion. Hence, the hypercalciuria and negative calcium balance. In contrast, those patients who had progressed to the inactive or equilibrium phase of their disease after several years had normal or slightly decreased bone turnover and were in zero calcium balance. It must be emphasized, of course, that these patients had extreme degrees of disuse osteoporosis at the time of study, despite active programs of active and passive physical therapy. The data were interpreted as indicating that the primary event in disuse osteoporosis is a significant increase in bone resorption and that, after loss of a substantial portion of the skeleton, a new equilibrium is established, with no further losses. This observation supports the concept proposed by Stubbs [15] that remodelling should not progress inexorably to total resorption of the skeleton despite persistent absence of normal stresses. It must be emphasized, however, that progression even to the point of the new steady state is an unacceptable risk for space explorers.

5. Prophylaxis

Based on the assumption that bed rest or enforced immobilization evokes a response simulating that of weightlessness, a number of studies have been performed and others are still in progress to evaluate potential prophylactic or therapeutic measures. Probably the most systematic of these studies are those initiated by McMillan, and now under the direction of Hulley, at the Public Health Service Hospital, San Francisco, California [27–30]. The program comprises complete bed rest of normal volunteer male subjects, with careful examination of multiple blood parameters and continuous measurement of calcium balance, along with bone densitometry. In consultation with Doctor Whedon and others, all experienced pioneers in this type of investigation, a variety of potentially beneficial agents has been tried. In Figs. 3 and 4 a portion of the data published by these investigators has been combined and re-plotted for comparison among the various agents. In both figures, values are shown as mean change in milligrams per day from ambulatory values prior to initiation of bed rest.

In Fig. 3 is shown the effect on urinary excretion of calcium. As demonstrated above (Fig. 2), the effect of bed rest without any specific prophylactic measures is a substantial increase in the urinary excretion of calcium, usually into the range considered as representing hypercalciuria. Despite vigorous exercise in bed by some subjects, there was no decrease in urinary calcium. Likewise, calcitonin (a hormone which specifically inhibits osteoclastic resorption

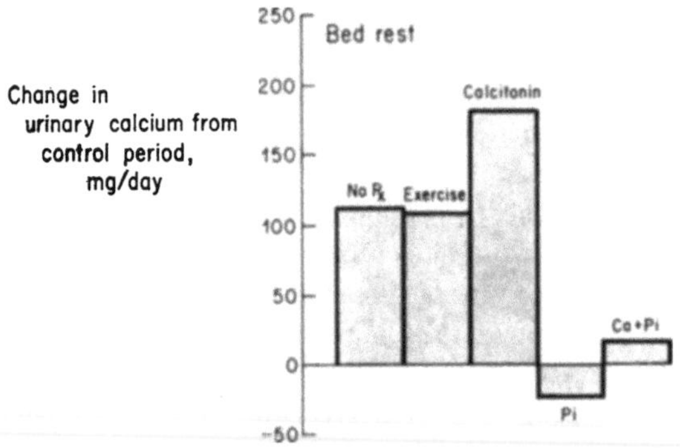

Fig. 3. Effect of various therapeutic agents on the urinary calcium during bed rest. Height of bar represents mean change (milligrams per day) from control values during normal ambulatory activity prior to bed rest. (Data collected from references [27–30]).

of bone) had no effect despite quite large doses. This may in part be due to the fact that calcitonin may have a direct renal tubular effect of increasing urinary excretion of calcium. In contrast to exercise or calcitonin, supplemental phosphate alone or in combination with a calcium supplement was extremely effective in reducing urinary calcium to or below control, ambulatory values. These data on phosphate are entirely consonant with our own previously reported observations [14]. One of the surprising observations noted in that report was that large amounts of calcium oxalate crystals appeared in the urine of subjects during immobilization, but none was observed in those subjects receiving supplemental phosphate. The importance of supplemental phosphate derives not only from its effect of inhibiting hypercalciuria, but also because it has been shown in a number of studies to inhibit nephrolithiasis in patients with "idiopathic stone disease", whether or not hypercalciuria was present [24]. Such studies have now been in progress for a number of years, and there has been no evidence of renal or other toxicity. In fact, a substantial number of patients have shown improved renal function, and one patient showed a disappearance of biopsy-proven nephrocalcinosis during therapy (L. H. Smith, personal communication).

5.1. Exercise

The effects of these same agents on calcium balance are shown in Fig. 4. Again, as noted previously, bed rest alone leads to significant calcium losses. Vigorous exercise was associated with no amelioration of these losses, the slight increase in loss probably not being significantly different from bed rest alone. It is perhaps not surprising that even vigorous exercise is unable to equal the stresses induced on bone by muscle action as a consequence of normal gravity. As pointed out by St. Clair Strange [31], the apparently simple maneuver of bending over a wash basin requires the lumbar erector spinae to exert a pull of at least three times the body weight, with corresponding stresses on the vertebrae. Similar, and even much greater, amplitudes of load are continuously being applied to most or all bones of the skeleton. No amount

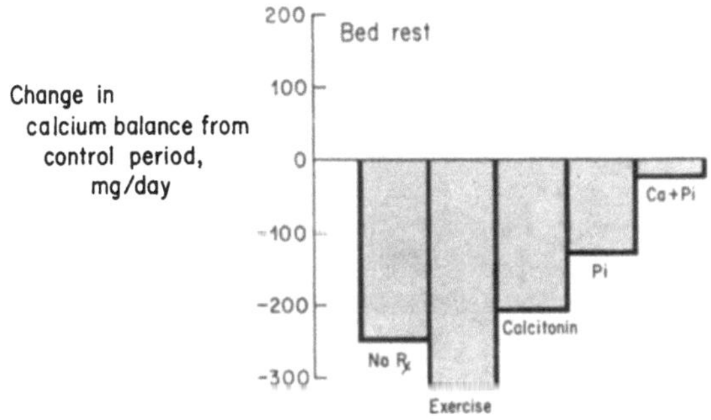

Fig. 4. Effect of various therapeutic agents on calcium balance during bed rest. Height of bar represents mean change (milligrams per day) from control values during normal ambulatory activity prior to bed rest. (Data collected from references [27–30]).

of exercise within the constraints of time and space available during space flight would seem to be capable of reproducing such stresses, and it is unreasonable to consider this a complete solution to avoidance of potential skeletal problems. In contrast, of course, the importance of exercise to maintenance of muscular tone and voluntary activity cannot be overemphasized. The recognition of "flabby" muscles by one of the astronauts following the flight of Gemini 7 [20] is reason enough to pursue exercise more vigorously than has been practised heretofore.

5.2. Calcitonin

Calcitonin was equally without effect on calcium balance despite the large doses employed. It is possible that the hypocalcemic effect of the calcitonin was sufficient to stimulate increased secretion of parathyroid hormone, as has been observed during therapy of Paget's disease, with subsequent counteraction of calcitonin action. It is also possible that osteocytic, rather than osteoclastic, osteolysis is the primary mechanism of immobilization osteoporosis and that this is not inhibited by calcitonin.

5.3. Supplemental Phosphate and Calcium

As previously noted [14], supplemental phosphate reduced negative calcium balance by about 50%. The addition of both calcium and phosphate, however, virtually abolished the negative calcium balance. (Although calcium supplementation alone may produce some reduction of calcium losses, it would not be satisfactory therapy because of the increased hypercalciuria which would be anticipated.)

It appears, therefore, that some of the alterations in calcium metabolism induced by bed rest may be prevented by appropriate supplementation with phosphate and calcium. It should be noted, in passing, that supplemental phosphate has the additional advantage of probably preventing hypercalcemia [14, 32]. The mechanism(s) by which phosphate produces these effects is unknown, although there are *in vitro* data which indicate both inhibition of resorption and stimulation of formation and calcification of bone, the latter probably predominating [33]. In addition, there are limited *in vivo* data suggesting that phosphate may stimulate synthesis of collagen [34]. Rasmussen [35] and Borle [36] by different techniques have demonstrated that phosphate facilitates the entry of calcium into cells, thus perhaps providing a basis for understanding the manner in which phosphate affects calcium transport and metabolism.

Unfortunately, the studies to date of phosphate (with or without calcium) supplementation provide scant evidence of its effect on skeletal architecture. Thus, although it may defend against total body loss of calcium, there is no evidence that it protects against architectural alterations which could still lead to structural weakening.

6. Summary

In summary, weightlessness imposes a unique kind of stress upon the skeleton, which in effect is the absence of normal compressive stresses. By analogy to bed rest or immobilization, this may lead to an alteration of piezoelectric impulses, which in turn may lead to an imbalance of the normal bone resorption/formation relationship, favoring increased resorption. Of prophylactic agents studied to date, only supplementation with calcium and phosphate shows promise of preventing the skeletal demineralization, but even this combination has not yet been demonstrated to maintain normal skeletal architecture. It is clear that more definitive techniques for studying this facet of skeletal metabolism must be developed and employed. Furthermore, additional studies during space flight are essential to determine the degree to which it conforms to the model of bed rest.

References

[1] J. H. SCOTT et al., J. Bone and Joint Surgery **39**, 134 (1957).
[2] G. H. BELL, Advancement of Science, September 1969 (p. 75).
[3] J. D. CURREY, Clin. Orthoped. and Related Res. No. 73, 210 (1970).

[4] J. Wolff, Das Gesetz der Transformation der Knochen, A. Hirschwold, Berlin 1892.
[5] C. A. L. Bassett, in: Calcified Tissues, Springer-Verlag, New York 1966 (p. 78).
[6] C. A. L. Bassett, Calc. Tiss. Res. **1**, 252 (1968).
[7] T. L. Jahn, Clin. Orthoped. and Related Res. **56**, 261 (1968).
[8] E. Jukada, J. Phys. Soc. Japan **12**, 1158 (1957).
[9] M. H. Shamos et al., Clin. Orthoped. **35**, 177 (1964).
[10] R. O. Becker, J. Arkansas Med. Soc. **62**, 404 (1965–66).
[11] J. E. Howard et al., Bull. Johns Hopkins Hosp. **77**, 291 (1945).
[12] J. E. Dietrick et al., Amer. J. Med. **4**, 3 (1948).
[13] R. P. Heaney, Amer. J. Med. **33**, 188 (1962).
[14] R. S. Goldsmith et al., Metabolism **18**, 349 (1969).
[15] D. Stubbs, Aerospace Med. **41**, 1126 (1970).
[16] G. P. Voss et al., Texas Rep. on Biol. and Med. **28**, 123 (1970).
[17] E. Israeli et al., Israel J. Med. Sci. **7**, 465 (1971).
[18] A. Iannaccone et al., Annals Int. Med. **52**, 570 (1960).
[19] J. H. Epstein, Amer. J. Med. **45**, 700 (1968).
[20] L. Lutwak et al., J. Clin. Endocrinol. and Metab. **29**, 1140 (1969).
[21] E. C. Reifenstein et al., J. Clin. Endocrinol. **5**, 367 (1945).
[22] D. Bronsky et al., Amer. J. Card. **7**, 833 (1961).
[23] J. E. Howard et al., Trans. Amer. Clin. Climat. Assn. **70**, 94 (1958).
[24] J. E. Howard et al., Amer. J. Med. **45**, 693 (1968).
[25] J. F. Aloia et al., Clin. Res. **19**, 365 (1971).
[26] P. B. Mack et al., Amer. J. Roentgenol. Radium Therap. Nucl. Med. **100**, 503 (1967).
[27] C. L. Donaldson et al., Metabolism **19**, 1071 (1970).
[28] S. B. Hulley et al., Clin. Res. **18**, 140 (1970).
[29] C. L. Donaldson et al., Clin. Res. **18**, 453 (1970).
[30] D. A. Hantman et al., Clin. Res. **19**, 476 (1971).
[31] F. G. St. Clair Strange, Brit. Med. J. **3**, 162 (1970).
[32] R. S. Goldsmith et al., New Engl. J. Med. **274**, 1 (1966).
[33] H. Rasmussen et al., Fed. Proc. **29**, 1190 (1970).
[34] R. S. Goldsmith et al., in: Phosphate et Métabolisme Phosphocalcique, D. J. Hioco, Laboratoires Sandoz, Paris 1971, p. 275.
[35] H. Rasmussen, in: Phosphate et Métabolisme Phosphocalcique, D. J. Hioco, Laboratoires Sandoz, Paris 1971, p. 7.
[36] A. Borle, in: Phosphate et Métabolisme Phosphocalcique, D. J. Hioco, Laboratoires Sandoz, Paris 1971, p. 29.

Life Sciences and Space Research X — Akademie-Verlag, Berlin 1972

REACTIVITY OF THE HUMAN BODY UNDER LONG-TERM HYPOKINETIC CONDITIONS

V. V. Parin and T. N. Krupina

USSR Academy of Sciences, Moscow, USSR

With an extension of manned space flights in the USSR and USA, the medical experts face the problem of human adaptation to long-term weightlessness and readaptation to 1 g.

The data obtained in orbital flights and post-flight examinations give evidence of functional changes in the physiological systems attributed mainly to effects of weightlessness. Present day techniques are inadequate to study the adaptation of the human body to extended weightlessness during space flight; hence, it is very important to carry out simulation experiments in earth-based laboratories.

We conducted a number of bed rest experiments of different duration and, as a result, came to certain conclusions concerning the origin and development of adaptive processes in various physiological systems. During a 120-day bed rest experiment the pattern and dynamics of functional changes were very close to those observed in shorter bed rest experiments. With an increase of hypokinetic exposures the reflex regulation of autonomic and somatic functions changed, i.e. the function of adaptive mechanisms that were under stress conditions reached a new level. This probably resulted in distinct clinical symptoms and syndromes. During long-term bed rest experiments, the rearrangement of regulation mechanisms followed certain stages; the amplitude of diurnal periodicity of functions gradually decreased thus indicating a step-by-step decrease of adaptive capabilities of the human body.

During the long-term bed rest experiment some of the functions were examined for the first time. For instance, investigations of eye hemodynamics demonstrated an increase of the intraoptic pressure (30–32 mmHg) by the 20th–30th hypokinetic day. An examination of the acoustic analyzer function showed an elevation of hearing thresholds during aerial conduction up to 25 dB, mainly at high frequencies. A study of blood coagulation indicated that all test subjects displayed an increase of blood thrombopoietic properties beginning from the 15th day and reaching a maximum by the 70th day. These changes in the blood coagulation system promoted prethrombotic developments in different parts of the vascular bed (in the region of the auricular artery and vessels of lower limbs). Investigations of lipid metabolism showed an increase of the content of total lipids, cholesterol and beta-lipoproteins. The changes of lipid metabolism were most pronounced after 35 days of exposure. These variations developed in parallel with changes in the blood coagulation and anticoagulation systems. The above developments may be of significance in the genesis of hypokinetic disturbances.

In our opinion, an inhibition of immunobiological and general reactivity of the human body (estimated by a decreased content of blood properdine, saliva and gastric juice lysozyme, and a reduced phagocytic activity of leucocytes) observed during the second–third hypokinetic months was the factor that caused allergic reactions in five out of ten test subjects in response to certain drugs administered for experimental purposes. Cytogenetic researches performed during the experiment gave evidence that the number of chromosome aberrations was within the limits of spontaneous variations of the level of chromosome aberrations.

The data accumulated are of interest for predicting possible changes in different systems during extended weightlessness effects. The possibility of preventing and treating hypokinetic disturbances with the aid of drugs seems very promising; e.g. the use of nerobol has yielded a good therapeutic effect with respect to fluid and calcium retention.

Life Sciences and Space Research X — Akademie-Verlag, Berlin 1972

ENERGY REQUIREMENTS OF MAN LIVING IN A WEIGHTLESS ENVIRONMENT

J. E. VANDERVEEN and T. H. ALLEN

USAF School of Aerospace Medicine (SME), Brooks AFB, Texas, USA

The ability to maintain energy balance is a vital factor in maintaining body composition. A negative energy balance requires that body tissue be consumed to sustain biochemical and physiological activity. Such a caloric imbalance coupled with reduced physical activity results in (among other things): a negative balance which can not be reversed by increased protein intake; negative balances for electrolytes; and a suspension of erythrocyte production. Body weight losses were experienced by all astronauts during Gemini and Apollo missions. Data on the magnitude of the changes, together with data on energy consumption, were used to calculate energy imbalances. These data, when compared with results obtained from precise energy balance measurements made on 64 men living in low pressure chambers, show close correlation. When energy requirements are expressed in kilocalories per kilogram of body weight, the difference in energy requirements among the astronauts and chamber subjects was small and not statistically significant. These data indicate that reliable prediction of energy needs for astronauts, during long-term space missions, can be made by studying either the astronauts or healthy subjects in a ground-based environment similar to that of the spacecraft. These data also indicate that changes in body weight and certain other body measurements detected during Gemini and Apollo missions were probably caused, at least in part, by a calorie deficit.

1. Introduction

The energy requirements of man while living in a space environment have been the subject of much speculation. It seemed logical to assume that living in a weightless environment would require less energy than on earth, since the work associated with counteracting the force of gravity would be eliminated. Research by Benedict and Musschhauser [1] reported in 1915 and later work by others [2] demonstrated that man's energy expenditure was greater while sitting than it was while lying down and greater for standing than for sitting. These data supported the supposition concerning man's energy expenditure in a weightless environment; however, the differences reported for sitting or standing versus lying down were small. For example the metabolic cost of lying in bed ranged 1.17–1.26 kcal min^{-1} for a 60 kilogram man whereas the costs for sitting and standing were 1.29–1.53 and 1.50–1.96 kcal min^{-1} respectively. By contrast the metabolic cost for walking, dressing and crawling ranged from 3.11–4.08, 3.20–3.40 and 4.50–6.66 kcal min^{-1} respectively. With the types of activity performed by the astronauts in the present space vehicles it becomes difficult to assess whether the calorie requirements would be substantially different from performing these same tasks on the surface of the earth.

For long-term space flights, the ability to estimate calorie requirements of the individual astronaut is critical. Insufficient energy in the diet results in the catabolism of body tissues including fat, protein and other organic constituents. If permitted to continue long enough, the losses in body tissues will result in decrements in physical performance and mental acuity. Excess energy in the diet is stored only as fat or glycogen. Although the consequences of fat storage are less serious, obesity can reduce physical performance and create problems for fitting of protective equipment such as pressure suits. In addition, food in excess to needs represents a logistics burden which could decrease the efficiency and reliability of the mission. Direct measurement of precise energy requirements during space missions has not been practical; however, a reasonable estimate of the energy requirements can be obtained from careful measurements of energy balance during each mission and correlation of these data with losses or gains in body tissues. Data obtained from the Gemini and Apollo missions, although incomplete, provide some insight on the astronaut's calorie requirements in space. This paper attempts to compare these data with data obtained from metabolic balance studies accomplished on healthy men living in low pressure simulators and consuming foods used in space feeding systems.

2. Experimental Procedure

Energy balances were performed in conjunction with a series of studies conducted to evaluate foods designed for space feeding systems. Accomplishment of these balances involved the precise measurement of the energy intake and the loss in the feces and urine. Aliquots of each food item were taken daily and combined into a composite sample. Individual samples were dried in a freeze dryer and ground in a Wiley mill until particles passed through a 60-mesh screen. Food samples which had a high fat content were mixed with dry ice to facilitate grinding. Fecal specimens were dried in a forced air oven at 57 °C until repeated daily weighings varied less than 1%. Dried specimens were combined into a four-day sample which was ground in a Wiley mill until the particles passed though a 60-mesh screen. Urine samples were pooled for a four-day period and a 50-milliliter portion was dried in a freeze dryer and held over sulfuric acid for subsequent analysis. Gross energy values for all foods and biological samples were determined by use of an oxygen bomb calorimeter. By use of these energy values, the digestibility and the metabolizability of food energy were calculated.

Subjects for this research were young military men carefully selected to match, except for age, as nearly as possible the astronaut population. Prior to and immediately following a study, each subject was given an extensive mental and physical examination to establish his state of health and detect possible changes created by the study. Part of this evaluation was the body composition of each subject as calculated using data obtained from body volume, body weight and total body water measurements [3]. By utilizing changes in body composition, energy balances were performed for the entire study. A series of 12 metabolic studies involving 40 men was accomplished. Twenty of these men lived in a low pressure simulator for periods of 16–32 days with an atmosphere of 70% oxygen and 30% helium at a total pressure of 258 torr. The remaining twenty men lived in a small metabolic ward. A second series of metabolic studies was conducted involving 24 men. These subjects consumed a formula diet at an approximate rate of 41 kilocalories per kilogram of lean body mass for 12 days and then consumed the same diet at 13 kilocalories per kilogram of lean body mass for 12 more days. Twelve of these subjects spent the 24 days of the study in a low pressure chamber with an atmosphere of 70% oxygen and 30% helium at 258 torr. The remaining twelve lived in a metabolic ward for the 24 days of the study.

Data for astronaut gross energy consumption and body changes were obtained from the literature [4–6]. Metabolizable energy consumption of the astronauts was calculated by using gross energy values of food consumed and multiplying the total by 0.95, the average value for metabolizable energy obtained from metabolic balance studies described above. Body tissues losses were calculated for all astronauts by averaging successive body weights obtained prior to the mission and subtracting the body weight measured 24 hours post recovery. It was assumed that any body water deficit would have been adjusted and that any change in body mass between the pre-flight weight and the recovery +24 hour weight represented a tissue loss.

3. Results

Data from the metabolic balance studies were compared in various ways to detect the relationship in body tissue changes and level of metabolizable energy supplied to the subject. In Fig. 1 the change in body mass of the subject is

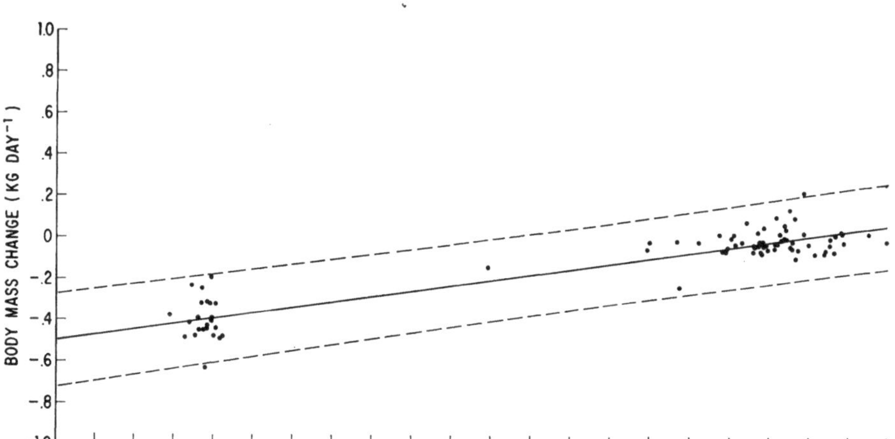

Fig. 1. Change in body mass in relation to metabolizable energy supplied to lean body mass.

shown in relation to the metabolizable energy supplied per kilogram of lean body mass. An average level of 41 kilocalories per kilogram of lean body mass was required to maintain body weight. The slope of the regression line is 0.01341 with a correlation of 0.9199. The relationship of change in lean body mass of the subjects to the metabolizable energy supplied per kilogram of lean body

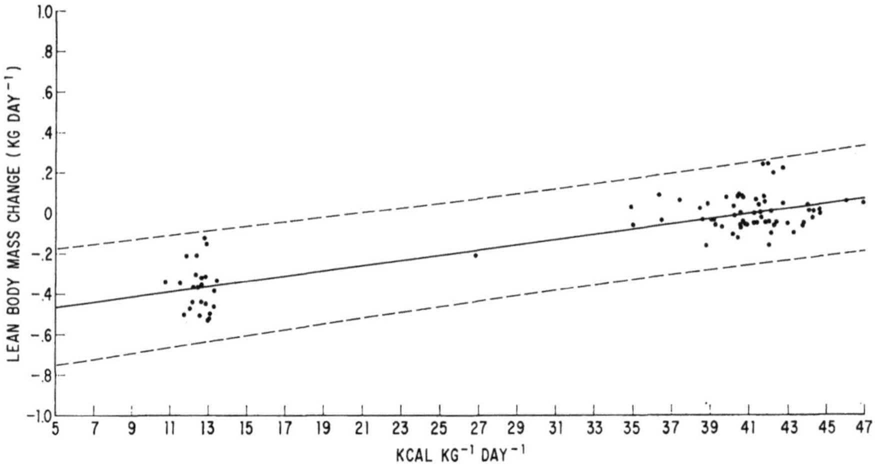

Fig. 2. Changes in lean body mass in relation to metabolizable energy supplied to lean body mass.

mass is shown in Fig. 2. The slope of the regression line is 0.01285 with a correlation of 0.8692. In Fig. 3 the change in body fat mass of the subjects is shown in relationship to the metabolizable energy supplied per kilogram of lean body

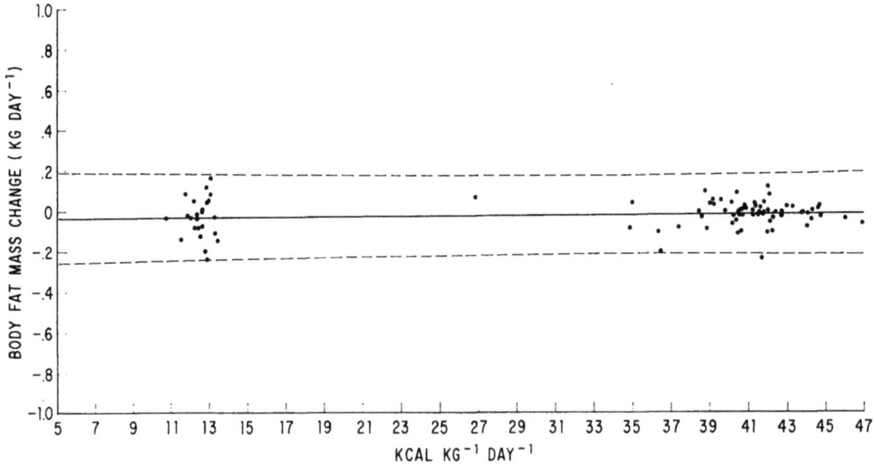

Fig. 3. Change in body fat mass in relation to metabolizable energy supplied to lean body mass.

mass. The slope of the regression line is 0.0006178 with a correlation of 0.1081, which is not significant. The relationship of change in body mass of the subject to the metabolizable energy supplied per kilogram of body mass is shown in Fig. 4. An average level of 37 kilocalories per kilogram of body mass was required to maintain body weight. The slope of the regression line is 0.01527 with a correlation of 0.9157. In Fig. 5 the change in body lean mass of the subjects is shown in relationship to metabolizable energy supplied per kilogram

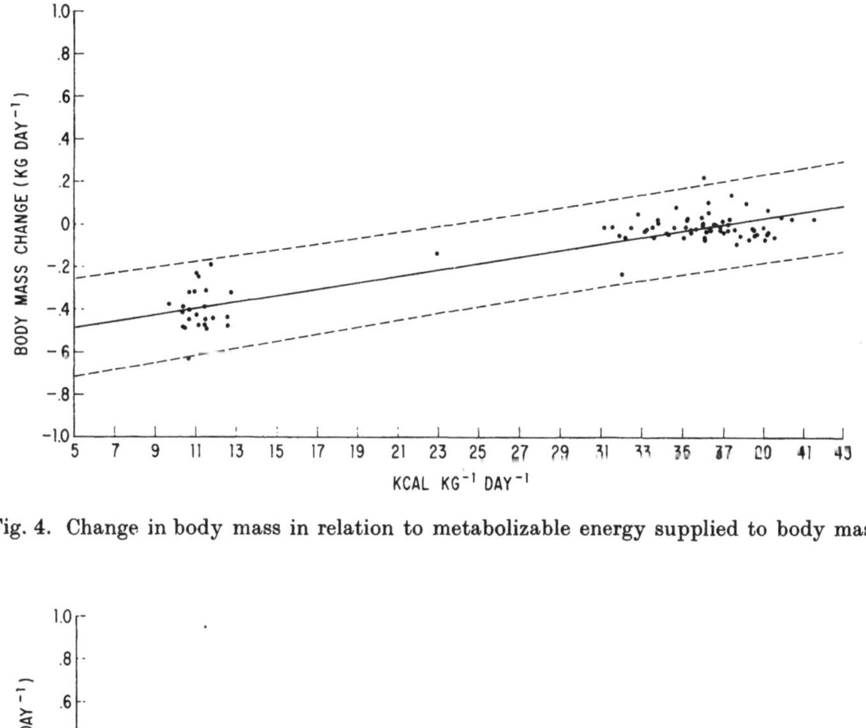

Fig. 4. Change in body mass in relation to metabolizable energy supplied to body mass.

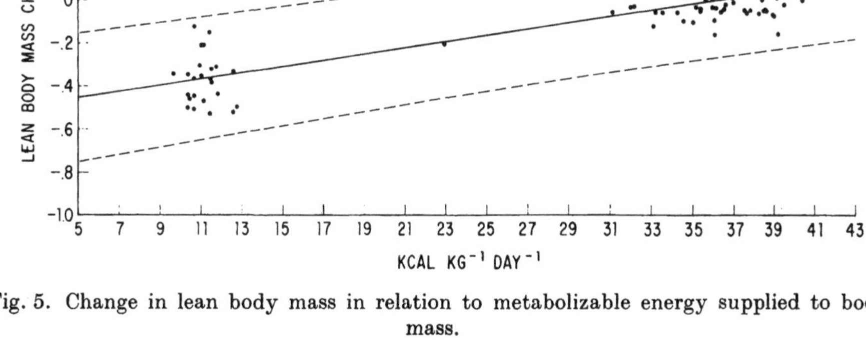

Fig. 5. Change in lean body mass in relation to metabolizable energy supplied to body mass.

of body mass. The slope of the regression line is 0.01450 with a correlation of 0.8569. The relationship of the change in body fat mass of the subjects to the metabolizable energy supplied per kilogram of body weight is shown in Fig. 6. The slope of the regression line is 0.00084 with a correlation of 0.1284, which is not significant. In Fig. 7 the change in the body mass of the astronauts is shown in relation to the estimated metabolizable energy supplied per kilogram of body weight. The slope of this regression line is 0.0196 with a correlation of 0.5624. An average of 32 kilocalories per kilogram of body mass was required to maintain body weight.

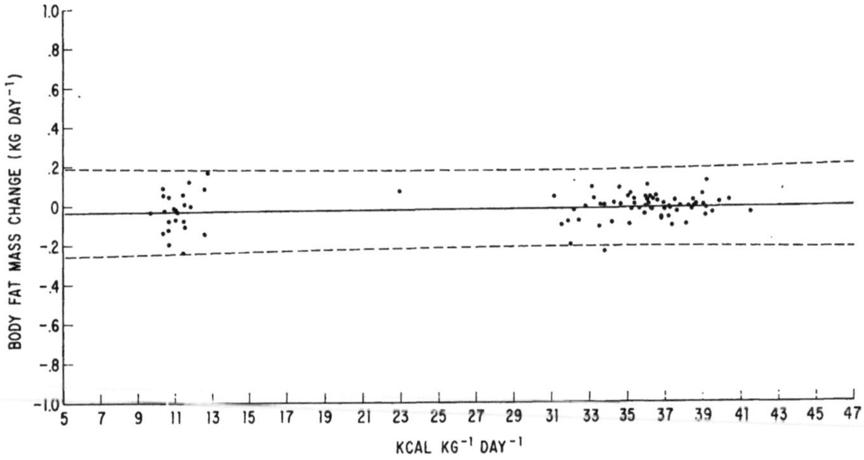

Fig. 6. Change in body fat mass in relation to metabolizable energy supplied to body mass.

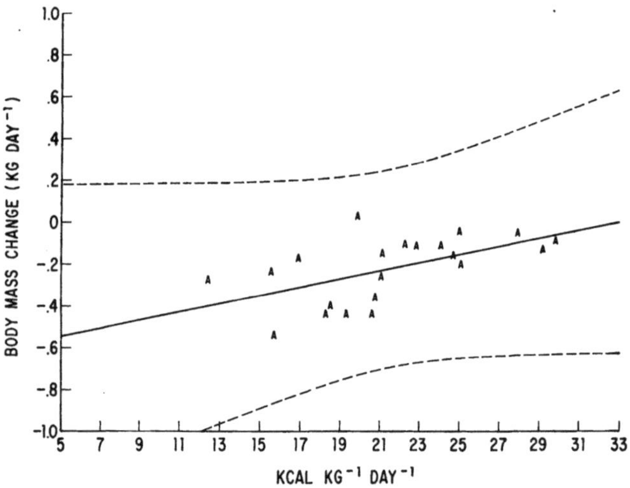

Fig. 7. Change in astronaut body mass in relation to metabolizable energy supplied to body mass.

4. Discussion

Several methods of estimating calorie requirements of humans have been proposed using functions of body weight with and without functions of body height [7, 8]. A study of these proposed formulas reveals that the accuracy of the predictions becomes less with extremes in body size and percentage of body fat. This condition is related to the fact that a considerable part of the body mass makes little or no direct contribution to the basal metabolism or to the energy transformations that make up the total calorie expenditure. With the exception of fat, however, the non-contributing mass remains in relative proportions to the living protoplasm which is considered the site of energy transformations. Based on this concept, changes in body tissue mass were compared with energy supplied per kilogram of lean body mass. The comparison of body mass changes of the subjects with the base of energy supply shows a high correlation. The comparison of body mass changes of the subjects with the energy supplied per kilogram of body mass was also highly significant with an almost identical correlation. The fact that the body composition of the subjects was within a narrow range as a result of selection is probably the reason for close correlation between the two comparisons. Body composition measurements were not performed on American astronauts prior to and immediately after flight, therefore, the only comparison that can be made is that using body weight. Since the astronauts were in excellent body condition prior to the flight, it can be assumed that body fat levels are also within a low narrow range. Therefore a comparison of data obtained from the subjects used in the metabolic studies and the astronauts during the space flights is valid. A review of the data in Fig. 4 and 7 shows a 5 kilocalorie per kilogram of body weight difference in the energy level necessary to maintain body weight for the astronauts during space flight and the subjects on earth. In previous studies [9, 10] the calorie levels for men living in space simulators were 32.7 and 34.0 kilocalories per kilogram of body weight, respectively. The low level of correlation for the available astronaut data limits the reliability of predicting precise energy requirements. As indicated in experimental procedure, it was assumed that body water imbalances would have been adjusted within 24 hours. Since body water measurements were not made there is no assurance that such an adjustment occurred. Such an error would result in change of slope for the regression line. Based on the data obtained from the metabolic balance studies, it is highly desirable that body composition measurements be made on all astronauts prior to and immediately following each future space flight. The high correlation between body lean mass changes and the level of metabolizable energy supplied and the low correlation between body fat mass changes and the level of metabolizable energy supplied indicates body lean mass was catabolized in preference to body fat mass in these subjects. The preferential catabolism of body lean mass was possibly influenced by the low level of body fat and the reduced physical activity of the subjects. Since the body composition of the astronauts can be assumed to be similar to that of the subjects used in this study, it is reasonable to suspect that the astronauts also catabolized body lean mass to meet energy needs not supplied by the diet.

Acknowledgments

The authors wish to recognise Mr. Clarence F. Theis for the accomplishment of body volume, body mass and residual lung volume measurements; and Mr. Lester L. Dziuk and Mr. Frank Bernal for the performance of energy determinations.

References

[1] F. G. BENEDICT and H. MUSSCHHAUSER, Carnegie Institution Publication No. 231 (1915).
[2] C. F. CONSOLAZIO, R. E. JOHNSON and L. J. PECORA, Physiological Measurements of Metabolic Functions in Man, McGraw-Hill, New York 1963 (p. 328).
[3] T. H. ALLEN, Aerospace Med. **34**, 907 (1963).
[4] C. A. BERRY, Aerospace Med. **41**, 500 (1970).
[5] P. A. LACHANCE and C. A. BERRY, Nutrition Today **2**, 2 (1967).
[6] P. A. LACHANCE, R. A. NANZ and M. V. KLICKA, NASA Tech. Mem. TMX-58010 (1967).
[7] Calorie Requirements, FAD Nutritional Studies No. 15 (1957).
[8] Recommended Dietary Allowances, National Academy of Sciences Publ. No. 1694, p. 1 (1968).
[9] B. E. WELCH, NASA Spec. Publ. No. 70, p. 181 (1964).
[10] V. D. BYCHKOV and P. P. IVANOV, Kosm. Biolog. i Med. **3**, 58 (1969).

Life Sciences and Space Research X — Akademie-Verlag, Berlin 1972

INFLUENCE OF COSMOS 368 SPACE FLIGHT CONDITIONS ON RADIATION EFFECTS IN YEASTS, HYDROGEN BACTERIA AND SEEDS OF LETTUCE AND PEA

Yu. G. Grigoryev, V. P. Benevolensky, Yu. P. Druzhinin, Yu. I. Shidarov, V. I. Korogodin, L. V. Nevzgodina, A. T. Miller and L. S. Tsarapkin

Institute of Medical and Biological Problems, Moscow, USSR

Radiobiological studies have been carried out on board the Cosmos 368 satellite, launched on 8 October 1970 and returning to earth on 14 October 1970. Yeast diploid cells *Saccharomyces ellipsoides*, Megri strain 139-B, haploid cells *Zygosaccharomyces Baili*, hydrogen bacteria *Hydrogenomonas eutropha*, strain Z-1, Berlin variety lettuce seeds and Capital variety pea seeds were used in these experiments.

The biological specimens were irradiated with gamma-rays at dose rates of 71.8 and 6.7 rad d^{-1} suring the packaging of containers and after returning the samples to the laboratory. It was found that both on pre-radiation and post-radiation exposure space flight factors did not greatly influence radiobiological effects.

One of the main purposes of the radiobiological experiments on board artificial earth satellites is the study of the modifying influence of weightlessness and other non-radiation space flight factors on the biological effects of cosmic radiation. The general radiobiological findings and the available materials of post-flight radiobiological tests suggest that such an influence does exist but so far little can be said with confidence about the trends and mechanisms involved in radiosensitivity modifications under space flight conditions. The amount of information available on the combined effect of all space flight factors is still very limited and therefore it is difficult to imagine the possible consequences of radiation and non-radiation effects which may become manifest at various levels in biological organisms.

The events occurring at molecular and cellular levels are of great significance. Their consequences may result in disturbances of the most basic functions of living systems such as growth and cell division, stability of genetic apparatus, enzymic processes, cell membrane permeability, etc.

For studying this question investigations on the simple organisms, which are generally good objects in which to observe any modification of radiosensitivity, are of paramount importance. In this respect the radiobiological experiments conducted on man-made satellites having an on-board radiation source, similar to those tests carried out during the Biosatellite 2 mission, will have an important role. Experiments with pre- and post-flight irradiation of test specimens flown on the satellites launched into the orbits beyond the radiation belts during the quiet sun periods aid in studying the effects of non-radiation factors on radiosensitivity. Such a scheme of experiments, their relative simplicity, the great possibility of varying radiation conditions, and the low costs, enable one to examine separately the influence of the complex of non-radiation factors on the various stages of the formation and development of radiation damage.

The time discrepancy between the action of radiation and that of non-radiation space flight factors in such an experimental scheme does not invalidate the tests. Basically, it is unlikely that the space flight environment will affect the initial mechanisms of energy exchange and the primary physico-chemical reactions proceeding in an irradiated organism for a short period of time. It appears that the influence of a combination of space flight factors will manifest itself in successively later stages of the radiation injury directly associated with physiological and metabolic processes. Disturbances before or after irradiation due to exposure to non-radiation space flight factors must affect the radiosensitivity of the organism. Thus, despite the time discrepancy between the action of radiation and non-radiation space flight factors, the study of the influence of the latter on radiosensitivity upon a given scheme of experiments is theoretically quite warranted. Furthermore, as is known, exposure to the most dangerous component of cosmic radiation — solar flare protons — is not constant. In such a situation the modification of radiation damage might be expected to be first of all on the exposure to the space flight factors before and after irradiation. The experiments with pre- and post-flight irradiation of the biological objects thus simulate to some extent the actual radiation environment of space flight.

Proceeding from these premises radiobiological studies were carried out on board the Cosmos 368 satellite. The satellite was launched on 8 October 1970 and placed in a near-earth orbit.

Yeast diploid cells *Saccharomyces ellipsoides*, Megri strain 139-B, haploid cells *Zygosaccharomyces Baili*, hydrogen bacteria *Hydrogenomonas eutropha*, strain Z-1, Berlin variety lettuce seeds and Capital variety pea seeds were used in these experiments. All these specimens except the hydrogen bacteria were carefully studied from the radiological point of view in previous ground-based tests. On board the satellites the specimens were in small metal containers, of which a general view is given in Fig. 1. The ampoules and the small glass Petri dishes were placed in a large container; the four smaller packages contained the seeds of plants. To counteract drops in temperature the internal surface of these containers was coated with a foam plastic, "porolon" (for seeds) or with foam polyurethane (for micro-organisms). Besides the biological material each container had an assembly of several thermoluminescence dosimeters. The containers on board the satellite were placed in different sites. Temperature at these sites ranged from +19 to +23 °C, but in the descent stage of the flight it increased to +35 °C for a short time. The pressure inside the capsule was between 760 and 820 mmHg. The initial atmosphere consisted of 40% O_2 and 15% CO_2. The satellite returned to earth on 14 October 1970; thus the flight lasted six days. Half an hour after retrieval of the satellite the containers with the test specimens were removed from the capsule and on the next day the samples were delivered to the laboratory.

In accordance with the requirements of the experiment part of the biological material was put into containers 12 days before and part just prior to placing

Fig. 1.

the containers aboard the satellite. After housing aboard the satellite the samples were maintained at the launch site for three days at room temperature. To fulfil such conditions it is necessary to use only objects which can remain in a dormant state (seeds of plants, cells in a stationary phase of growth) or in a state of partially suppressed metabolism induced by nutrient medium impoverishment and limitations of air oxygen diffusion (aqueous suspension of micro-organisms in ampoules). Before being placed aboard the satellite the seeds of plants were stored at room temperature, and micro-organisms maintained at the temperature of thawing ice.

In parallel with the main experiment there were similar containers with duplicates of the specimens which, together with the test samples, were a so-called "transport control" up to the time of placing the containers aboard the capsule. Later, these specimens were returned to the laboratory where they were maintained at a temperature close to that of the spacecraft. Five days later a synchronous experiment with plant seeds was performed in a special chamber simulating the space flight environmental conditions, i.e. temperature, pressure, humidity and cabin atmosphere composition. In the case of micro-organisms only temperature profiles were simulated under laboratory conditions, since all specimens were almost completely isolated from the ambient air.

The biological specimens were irradiated with gamma-rays at dose rates of 71.8 and 6.7 rad s^{-1} during the packaging of the containers and after returning the samples to the laboratory. The radiation damage was generally estimated on an arbitrary scale.

The main findings of these experiments are presented in the Table 1. A detailed description of the tests and the results of experiments performed on each biological specimen are contained in the papers that will be published in the *Journal of Space Biology and Medicine* (in Russian). However, all these papers

Table 1

Influence of Cosmos 368 satellite space flight conditions on radiation effects of yeasts, hydrogen bacteria and seeds of lettuce and pea

		Exposure	Pre-flight		Post-flight		
(1)	(2)	(3)	(4)	(5)	(6)	(7)	(8)
Yeast Saccharomyces ellipsoides	Cells in aqueous suspension	Inactivation to infinite proliferation (formation of normal colonies)	20	Slight decrease	20	None	
			40	None	40	Slight decrease	
			80	None	80	None	
			120	Slight decrease	120	None	
			160	None	160	None	
		Inactivation after 1–4 cycles of cell division	20	None	20	None	
			40	None	40	Slight decrease	
			80	None	80	Slight decrease	
			120	Slight decrease	120	None	
			160	None	160	None	
	Cells on agar	Inactivation to infinite proliferation	—	—	20	Slight decrease	
			—	—	40	None	
			—	—	80	None	
			—	—	120	None	
			—	—	160	None	
		Inactivation after 1–4 cycles of cell division	—	—	20	None	
			—	—	40	None	
			—	—	80	Decrease	
			—	—	120	Decrease	
			—	—	160	Decrease	
Yeast Zygosaccharomyces Baili	Cells on agar	Inactivation to infinite proliferation	—	—	1.5	None	
			—	—	3.0	None	
			—	—	6.0	None	
			—	—	10.0	None	
		Inactivation after 1–4 cycles of division	—	—	1.5	None	
			—	—	3.0	None	
			—	—	6.0	None	
			—	—	10.0	None	
Hydrogen bacteria	Cells in aqueous suspension	Inactivation to infinite proliferation	0.5	None	—	—	
			1.0	None	—	—	
			2.0	None	—	—	
			4.0	None	—	—	
			6.0	None	—	—	
Pea	Air-dry seeds	Chromosomal aberrations in seedling cells	5.0	None	—	—	
			10.0	None	—	—	
			10.0	None	—	—	
		Formation of cells with bridges	5.0	None	—	—	
			10.0	None	—	—	
		Formation of cells with fragments	5.0	None	—	—	
			10.0	None	—	—	

Table 1 (contd.)

(1)	(2)	(3)	(4)	(5)	(6)	(7)	(8)
Lettuce	Air-dry seeds	Chromosome aberrations in ana- and telophase	10.0	None	–		With previous storage of seeds for 6 days at 10 °C
		Number of aberrations per one aberrant cell	10.0	None	–	–	
		Chromosome aberrations in ana- and telophase	10.0	Increase	–	–	With previous storage of seeds for 6 days at 30 °C
		Number of aberrations per one aberrant cell	10.0	Slight increase	–	–	With previous storage of seeds for 6 days at 30 °C
Lettuce	Air-dry seeds	Stimulation and suppression of growth processes of aerial portion of plant	1.0	None	1.0	None	
			5.0	Increased suppression	5.0	None	
			10.0	Increased suppression	10.0	None	
		Stimulation and suppression of growth processes of subsurface portion of plant	1.0	None	1.0	Slightly increased stimulation	
			5.0	None	5.0		
			10.0	None	10.0		
		Activation and suppression of enzymic activity of vegetative plants					
		a) polyphenyloxidases	1.0	None	–	–	
			10.0	Suppressed activation	–	–	
		b) catalases	1.0	None	–	–	
			10.0	Increased suppression	–	–	
		c) peroxidases	1.0	None	–	–	
			10.0	None	–	–	
		Formation of abnormal metabolites in 30-days old plants	1.0	Decrease	1.0	Decrease	
			10.0	Increase	10.0	Increase	

(1) Organism; (2) experimental in-flight conditions; (3) radiation effect; (4), (6) exposure dose (krad); (5,) (7) space flight factor effects; (8) notes.

will deal with the overall effects of the dynamic space flight factors and weightlessness. Other effects, such as temperature conditions, changes in gas composition, pressure, humidity and transportation conditions are excluded to some extent by the corresponding control experiments. In the present case the influence of cosmic radiation and the action of a gamma-ray source, installed on board the satellite with the aim of studying its performance, can be ignored: the maximum dose registered inside the assemblies was only 0.7 rad; its effect is beyond the sensitivity of the methods used.

From the data obtained it was found that both on pre-radiation and post-radiation exposure, space flight factors did not greatly influence radiobiological effects. As most measurements showed, there was no modification of radiation damage; in cases where modification was present its magnitude was insignificant and it is probable that it was due to some factors not associated with the flight.

The efficiency of growth, seed germination, mitotic index in seedling cells etc., i.e. the indices that were unchanged at the selected radiation doses, were analysed in these experiments. After the flight these indices were also unalterd, that is, the space flight conditions did not contribute to revealing latent radiation damages.

In the experiment with yeast diploid cells an attempt was made to study the effect of space flight conditions on post-irradiation recovery. As is known, the yeasts are exceedingly good objects in which to investigate these phenomena. The influence of non-radiation space flight factors on post-irradiation recovery was studied both on non-irradiated cells and on irradiated cells in a weightless environment. No modification of recovery effect was found in any of these cases.

In summary, it may be concluded that radiosensitivity in dormant organisms with low metabolic activity is not changed at all or is changed insignificantly under the influence of the dynamic space flight factors and weightlessness (over six days). The question concerning the effects of specific space flight factors on the growth and active functioning of organisms is still unsettled. The findings of the biological experiments carried out during the Biosatellite 2 and Cosmos 110 satellite missions indicate that in such a situation the probability of enhancing or reducing the radiation effect is quite high, particularly in extended space flights.

Life Sciences and Space Research X — Akademie-Verlag, Berlin 1972

EFFECT OF WEIGHTLESSNESS ON CELLS AND TISSUES

B. F. Edwards

Emory University, Atlanta, Georgia, USA

The response of cells and tissues to weightlessness is demonstrated by reactions of the wheat seedling to both orbital and simulated weightlessness. Wheat seedling shoots and roots grown in Biosatellite 2 have been compared with those grown erect on earth and horizontally on the clinostat. True weightlessness increases root cell length (14–17%) and decreases the number of dividing cells (20–43%). Simulated weightlessness produces no significant changes in these measurements. Nuclear volume is increased 13–26% by true weightlessness, but no such increase results from simulated weightlessness, with one exception.

Root tip cells of *Triticum* have also been compared with those of *Tradescantia* orbited in the same vehicle (Marimuthu, Sparrow and Schairer). The starch grain distribution is similar in cells of both species. There is a significant number of abnormal nuclei in orbited roots, as well as fewer dividing cells. Clinostat simulated weightlessness does not produce similar changes in mitotic rate, cell length nor nuclear size and shape, although orientation and starch grain distribution are comparable.

Coleoptile and root tissues in longitudinal and cross section at light and electron microscopic levels illustrate the effects of weightlessness and show that on a cellular level, true weightlessness cannot be simulated perfectly by the clinostat on earth.

Life Sciences and Space Research X — Akademie-Verlag, Berlin 1972

OFO EXPERIMENTAL TECHNIQUES AND PRELIMINARY CONCLUSIONS: IS ARTIFICIAL GRAVITY NEEDED DURING PROLONGED WEIGHTLESSNESS?

T. Gualtierotti and F. Bracchi

Department of Physiology, University of Milan, Italy

The technique of single unit recording from body systems generating electrical pulses coherent with their basic function (CNS, muscles, sense organs) has been proved feasible during the OFO A orbital flight, an automatic physiological experiment. All microelectrode implants survived the lift off of a Scout vehicle. The far-reaching impact of such a technique in biological space research and in the laboratory is discussed.

The results of recording 155 hours of orbital flight of pulses from the nerve fibres of four vestibular gravity sensors in two bull frogs indicate that the vestibular organ adjusts to zero *g*. As all the other biological changes observed during orbit are due to lack of exercise, it is concluded that artificial gravity might not be necessary during prolonged space missions or on low gravity celestial bodies.

1. Introduction

It is generally agreed that the activity of the vestibular apparatus is of paramount importance in relation to the state of weightlessness achieved for an extended period of time in space flights. In 1964 Yazdovsky et al. [1] concluded "against the background of the altered afferent impulses, apparently the afferentation from the labyrinth of the inner ear can become predominant and this produces the air sickness syndrome". More recently Berry [2] stated "It appears that the opportunity to move about more freely in the Apollo cabin than in previous spacecraft is a factor producing the motion sickness problem. Sensory inputs from the semicircular canals to the central nervous system during head movements in space are thought to be enhanced due to altered activity of the otolith organs in the weightless state. This is a significant problem which must receive continued attention in the space program, for it can markedly affect astronaut performance". It is therefore obvious that a direct study of the response of the labyrinth to an extended period of weightlessness is a necessary step in the development of space biology. It is best to start the investigation of the vestibule at the single unit level as this will allow quantitative analysis; it is also more convenient to record from the primary neurones, as at the second order neurones level already the convergence of different inputs both from the other parts of the vestibule and from other afferents of the body will complicate analysis [3, 4].

Of the different sensors of the inner ear the most obvious choice is the stato-receptor, i.e. the one which responds to a steady position of the head (owing to the gravitational components) or to a steady linear acceleration, with a stationary average discharge; after some initial adaptation the response will assume constant characteristics for an indefinite period of time (Fig. 1). Therefore a recognizable different activity will correspond to each different position of the head. The purpose of the OFO A space project is to record the direct output of such receptors at 1 g and during an orbital flight at 10^{-3} g. The modulation of a spike train forms the information code of this output, along the nerve fibres that connect the receptor cells to the secondary neurones in

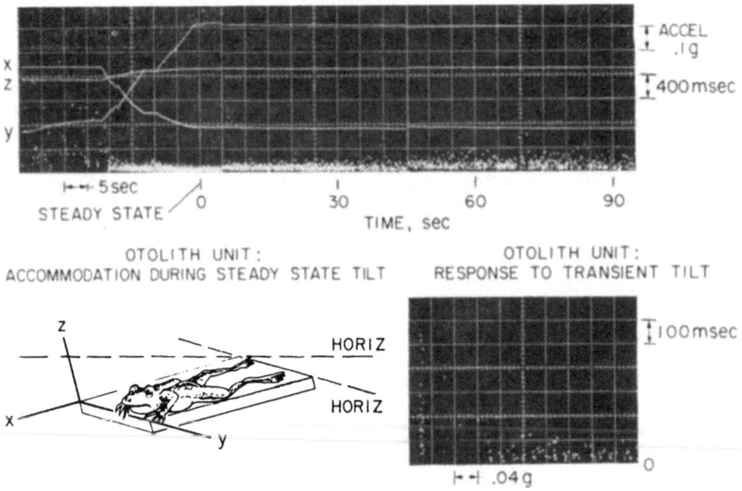

Fig. 1. Response to tilting of statoreceptor in the bull frog. In the top figure the continuous upper lines correspond to X, Y, Z accelerometers. The white dots correspond to the consecutive intervals measured as a distance from the bottom line of each dot. Note the decrease in the value of the average interspike interval during the actual tilting, showing some accommodation at steady position during the first 20 seconds; a stationary response is reached thereafter.
Bottom left: schematic of frog and direction of accelerations.
Bottom right: detail of response to increasing tilt: plot of the interspike interval shown as in the upper figure against increasing g.

the vestibular nuclei, or in the cerebellar centres directly. The significant information is the series of intervals between pulses: the durations of such intervals are a function of the input to the receptors, i.e. of the linear acceleration applied to the system. These receptors are therefore linear accelerometers: on earth the appropriate stimulus at the different steady head positions is the gravitational component in the direction of the functional axis of the sensors: such stimulus would be zero when the functional axis is situated at 90° from the vertical and equal to 1 g when coincident with the vertical. In between these two positions the stimulus will be proportional to the sine of the angle from the horizontal. As no receptor is able to cover the entire 90° solid angle but only a fraction of it, the overall coverage of the entire range of the 360° solid angle is assured by the fact that within the organ as a whole the sensors show different functional directions. The total information depends on two constants: the 1 g earth's gravity and the reciprocal position of the receptors in the macula.

If during an orbital flight the residual gravitational force or any undirectional linear acceleration in existence has an intensity below threshold for all or most of the receptors, then the vestibular statoreceptors will lack the principal stimulus necessary for their functions. However, (i) this does not mean absolute lack of stimulation, as for every movement of the head the corresponding acceleration will excite those receptors whose functional axes lie in the direction of the acceleration vectors; (ii) even at zero input the output would not be zero as all these cells show a spontaneous firing, namely a continuous volley of spikes as a basic activity of the cells independent of any input stimulation. The major change due to 0 g, as above defined, will be a disruption of the pattern of response in the multichannel system that constitutes the vestibular organ: here, for a given position of the head in the earth's gravitational field, every sensor will have a firing pattern corresponding to its situation with respect to the vertical: any movement will result in a response superimposed on this background systematic activity [5, 6]. In weightlessness this pattern will obviously disappear.

The importance of an eventual modification in the labyrinth activity results not only from the modification of its main function, the control of balance, but also from the fact that the vestibular organ is linked to other important functions of the organism like the muscular tone [7], the neurovegetative system [8, 9] (which explains the air sickness and motion sickness syndromes) and therefore the cardiovascular system, etc. The controlling action of the vestibule on the body seems to be exerted tonically through the spontaneous activity which is typical of all these receptors even at rest.

2. The Experiment OFO A

The OFO A mission was successfully launched on 9 November 1970 at 06 h GMT from the NASA centre at Wallops, Virginia, by means of a Scout B rocket and a satellite made to satisfy the requirement of a low g level. In two bull frogs the spike train data of four single fibres in the vestibular nerve corresponding to gravity sensors have been studied using a specially devised technique by means of chronic microelectrodes [10]. Recordings have been made at 1 g in the laboratory and during 155 hours orbital flight during which a maximum background acceleration of no more than 10^{-3} in all directions was maintained. Both during the tests on earth and in orbit the frogs were kept completely immersed in hyperoxygenated water (P_{O_2} 600 mmHg) at a constant temperature (16 $\pm$ 0.5) °C in a specially built capsule (FOEP): this capsule was composed of three main systems: (i) a life support system to ensure that the needs of the animal were met (Fig. 2), (ii) an amplifier system to increase the biological pulses to the level required by telemetry (Fig. 3) and (iii) a centrifuge that could be rotated up to a constant speed causing a centripetal acceleration

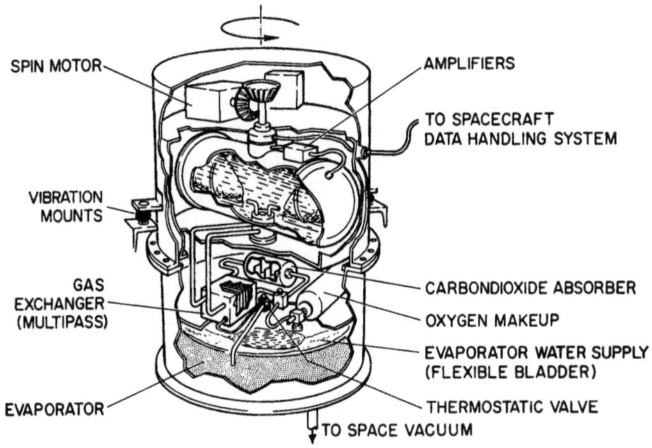

Fig. 2. Schematic of the FOEP showing the life supporting system including the lung (gas exchanger), the carbon dioxide absorber and the oxygen reservoir. The double circulation of water and oxygen is also evident, passing through the gas exchanger. The water filled centrifuge driven by the spin motor is also shown.

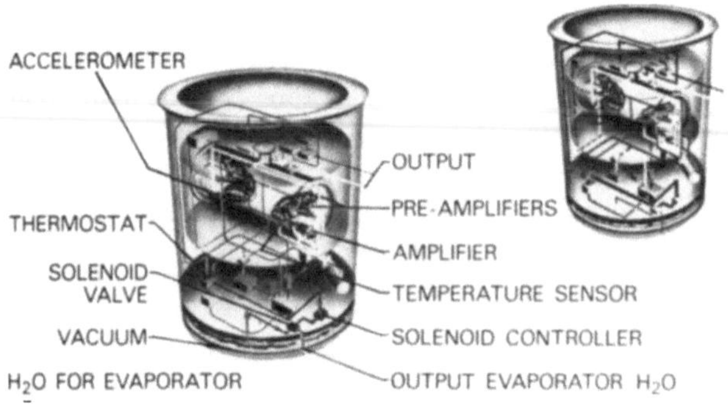

Fig. 3. Schematic of the electronics of the FOEP. The preamplifiers directly inbedded in the two bull frogs are shown one at each side of the head. The main amplifier raises the biological signals both from the otolith and the EKG to an output level adequate for telemetry. The temperature control system based on H_2O evaporation in vacuum is also shown.

of 0.6 *g* at the head of the frog (Fig. 2); in effect laboratory tests have shown that the centripetal acceleration evoked exactly the same responses from the gravitational receptors of the inner ear as did tilting on earth, which induced a gravitational component of the same *g* level in the same direction. The spike train data were recorded with minimum distortion by telemetry from six receiver stations distributed all over the surface of the earth under the orbit (Fig. 4). The EKG was continuously controlled as a vital index of the animal's condition as was the temperature and pressure of the water and the acceleration corresponding to the heads of the frogs, in its three components in space. The experiment was preplanned for the first 24 hours. At the end of this period

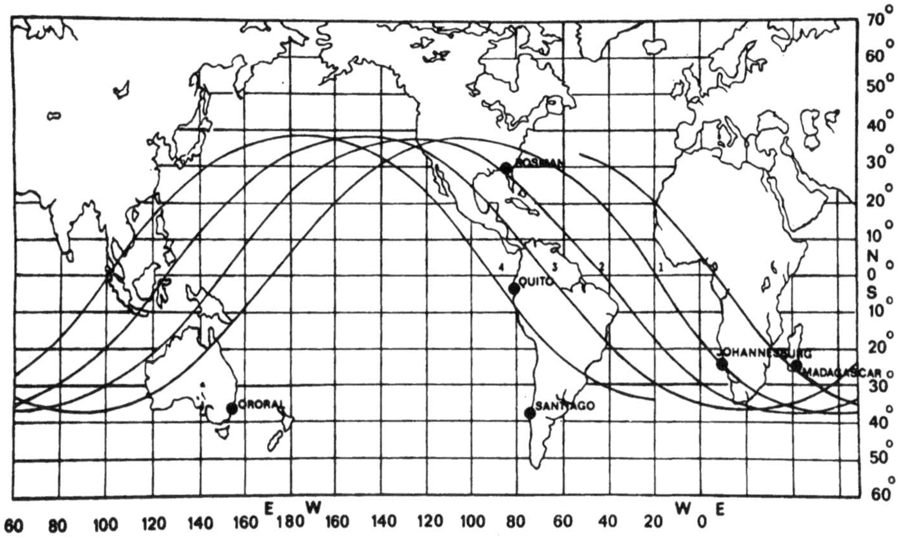

Fig. 4. Schematic of the distribution of receiver stations along the orbits.

it could be carried on according to one of three experimental routines chosen on the base of the preceding results (Fig. 5).

At the end of the experiment, which lasted 2 and a half times the officially established period (six and a half days instead of three: and it ended owing to the failure of the batteries on board) 300 effective hours of data were recorded: these have to be added to another 500 hours of data corresponding to control experiments and simulated mission on the ground. All the data were recorded on analog tape in the capsule and then delivered to the ground stations: a quick look analysis at Goddard Space Flight Center provided the necessary information during the flight to direct the continuation of the experiment: the time history of the sensor's firing and of the acceleration at the frog head (Fig. 6) was provided together with interspike intervals histograms (Fig. 7) and other pertinent statistics. At the end of the flight all data have been sent to Ames Research Center and to the Milan Physiological Laboratory and Space Biology Laboratory for analysis.

3. Results

Only partial results and partial conclusions have been reached so far: it has been calculated that a full analysis and definite conclusions of the results of the experiment will take from 18 months to two years to complete. No particular changes in respect of ground condition have been noted in the vital index of the frogs: the EKG showed the same characteristics as on the ground: probably submersion in water was responsible for minimizing the impact of the lift off; it is relevant to report that it was possible to record the EKG and the nervous activity through the entire procedure of launching and injection into orbit.

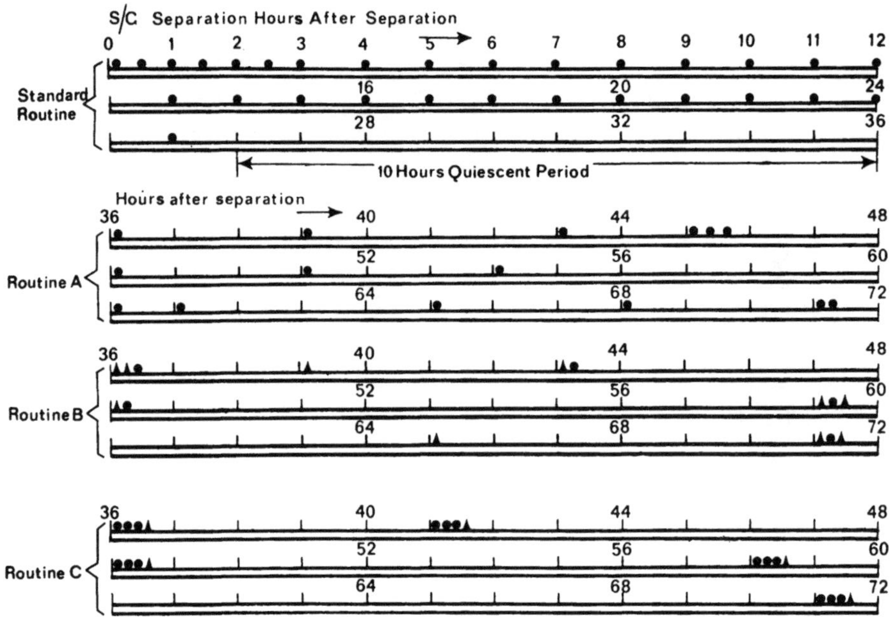

Fig. 5. Experiment cycles. ● standard 8-min centrifuge cycle, ▲ 8 min recording without rotation.
The centrifuge cycle started with $1^1/_2$ min of recording of data at rest followed by an increased rotation speed up to point 0.6 g in 8 s followed by 8 s at constant speed (0.6 g). The centrifuge then slowed down to zero speed in 14 s linearly. This was followed by $6^1/_2$ min of post rotatory recording of the otolith activity. At 2 min from the end of the centrifuge rotation a test calibration was performed for 5 s. Routines A, B and C followed the first 46 hours: Routine A was a continuation of the standard routine of the 24 hours assuming that no clear-cut results were detected in the first part of the mission. Routine B was based on the main alterations resulting in the activity following the centrifuge cycles (two or more immediately consecutive rotations were performed according to this plan). Routine C was based on the main alterations involving the spontaneous activity at rest. Consequently several consecutive 8 min recordings without rotation were performed. As a result of the quick analysis of the first 24 hours data routines B and C were used alternately for the continuation of the flight.

The nerve pulses were recorded by telemetry with minimum distortion (Fig. 7). They maintain the same parameter throughout the flight so that it was possible to be sure that the same unit was studied for the entire duration, and that no ill effect took place at the pick-up of the micro-electrode during the high acceleration (10 g) and vibration of the firing of the four stages.

The most important change that has been observed so far was in the spontaneous activity at rest. Fig. 8 shows diagrammatically such a change: during the first nine hours and up to the 46th hour the firing at rest slowed down so that the average interspike interval was more than four times longer in duration than on the ground. This phenomenon was followed by a rebound effect starting around the 48th hour and reaching its climax after the 72nd hour with a spontaneous firing twice as fast as the one shown by the same unit at 1 g. Recovery

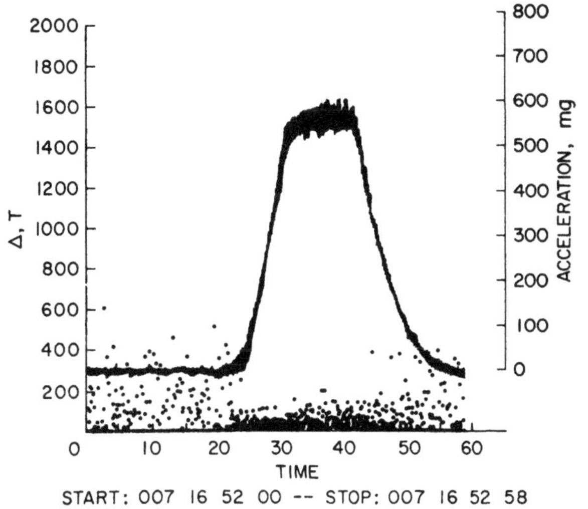

Fig. 6. Time history of the interspike interval changes and corresponding centripetal acceleration during one centrifuge cycle, otolith 4. Upper continuous trace: centripetal acceleration in g (ordinate on right). Black dots measure intervals as in Fig. 1 (ordinate on left).

to normal started after 120 hours. In Fig. 8 the typical behaviour of one representative unit is shown, the others followed the same pattern although with some shift in time. For instance the unit in Fig. 7 showed a slowing of firing rate at rest at the 39th hour and an increased rate after 120 hours. These changes are very significant: a large number of simulated missions on the ground had shown that for a comparable time of 5–6 days the vestibular unit in a FOEP identical to the one used in OFO A showed a constant firing rate within the normal standard deviation (Fig. 7).

The response to the centrifuge stimulation was not impaired significantly up to the third day: then for the following two days it showed a moderate decrease going back to normal on the fifth day (Fig. 8). Even this phenomenon showed some shift in time in the other units (Fig. 7) although the general pattern remained the same. The main consequence of the increased firing rate at rest and the decreased response to 0.6 g stimulation was that the unit was not able to distinguish between zero input and maximum input during this period: i.e. the receptors were not functioning. A trend to normality followed this stage of maximum alteration: unit No. 4 was practically normal after 120 hours (Fig. 8). All abnormalities during the flight were accompanied by a large increase in the variability of the firing rate as shown by the remarkable increase in the standard deviation.

During the flight some failures were observed: the most important ones (Fig. 8) were: (i) There was a critical decrease in the environmental temperature, from the 12th to the 17th hour, owing to the lack of response of the temperature control unit to a ground command. The low temperature reached at the 14th

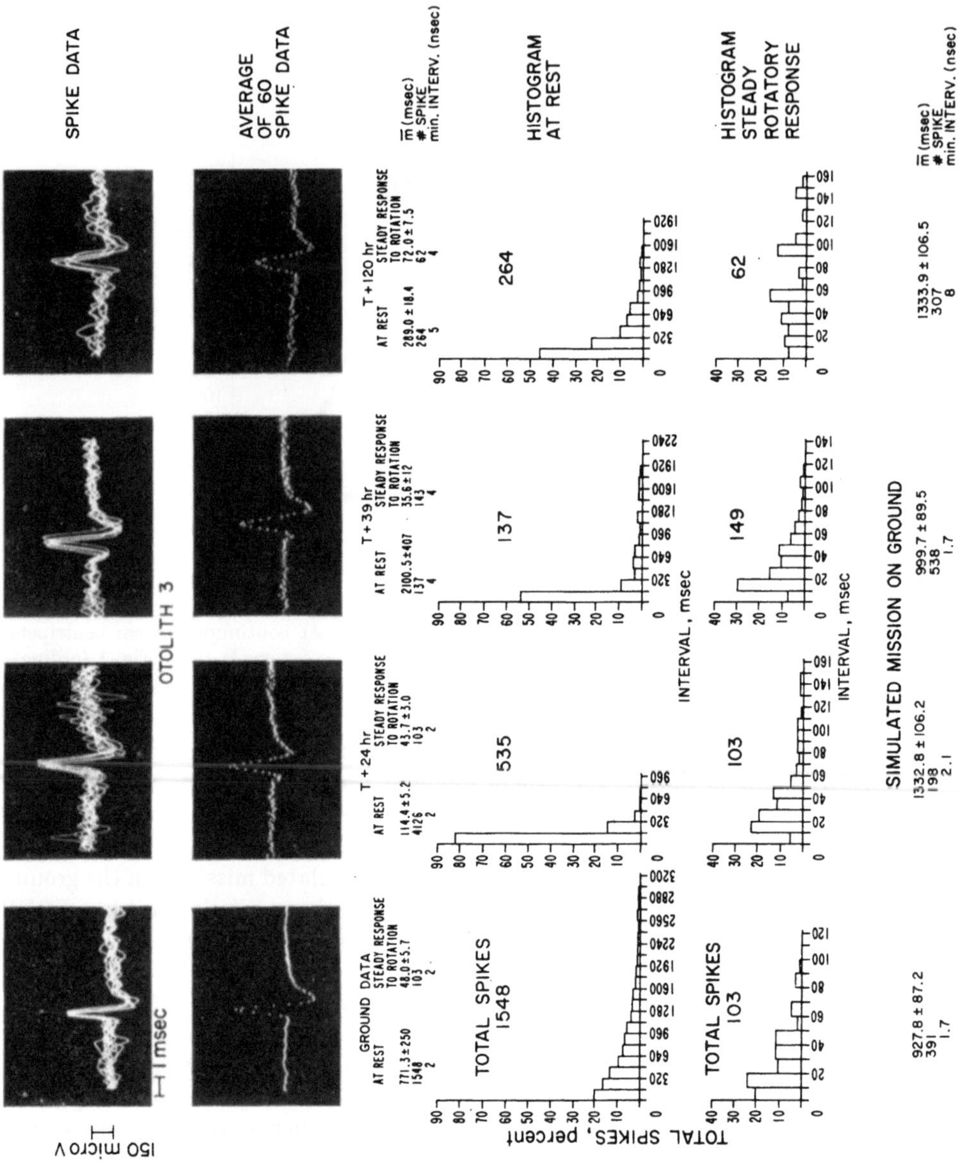

Fig. 7 Otolith No 3. Upper trace: Sample of spikes data recorded on the ground and thoughout the mission. Lower trace: Computer generated spikes as an average of 60 spike data. No significant changes in the spike parameter have been found as a result of 120 hours exposure to weightlessness. Also it is evident that the recording of data belongs to the same unit throughout the flight. Note also the slight increase of the noise level due to telemetry. The numerical values on the bottom of these figures correspond to statistics of the interval changes as represented by the corresponding histograms (bottom trace) at rest and during stationary response to steady rotation: for details on the relevant changes see text.

Under the histograms the results of one typical simulated mission on the ground performed using an identical FOEP as for the flight are reported for the same period of time, limited to the activity at rest as this is the one that changes more clearly.

The data of the mission on the ground indicate no statistically significant changes for the equivalent time of the space mission.

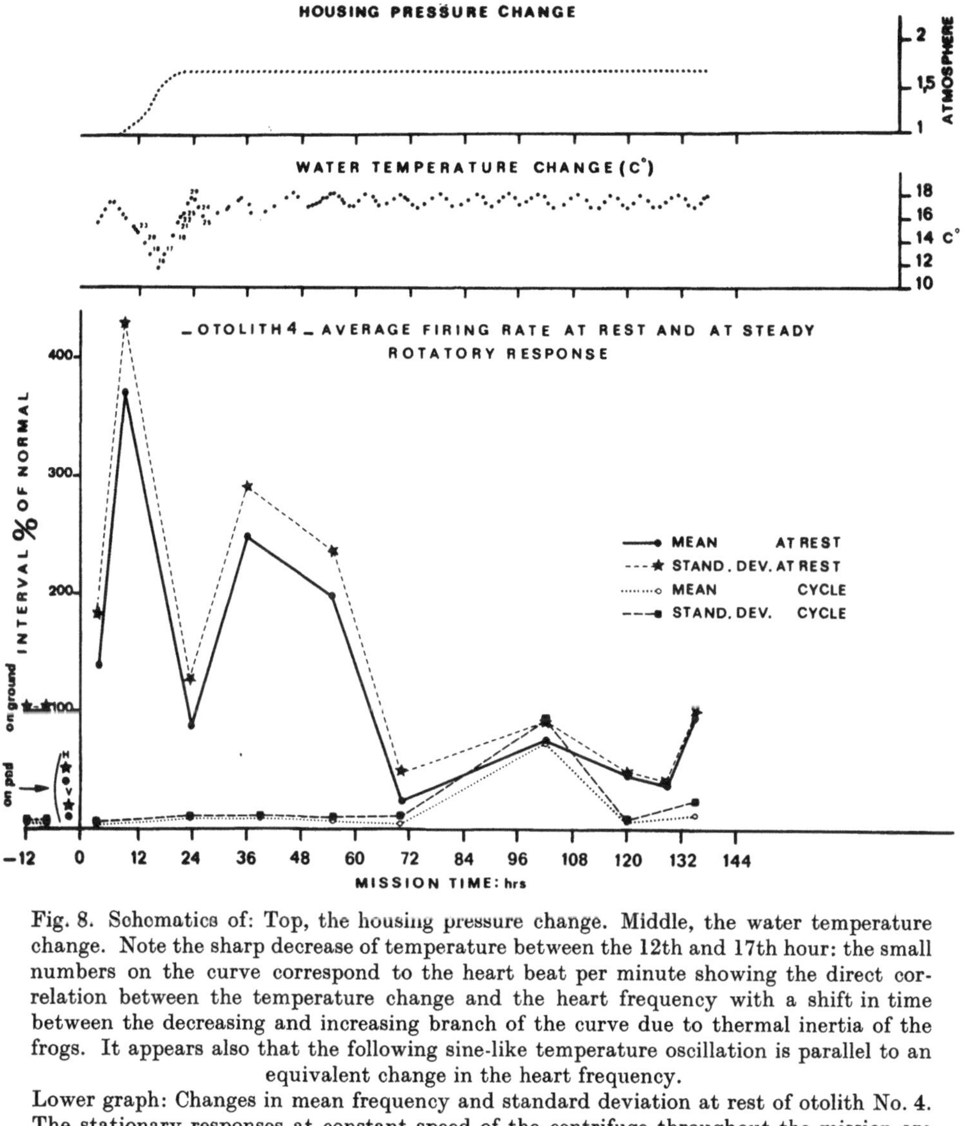

Fig. 8. Schematics of: Top, the housing pressure change. Middle, the water temperature change. Note the sharp decrease of temperature between the 12th and 17th hour: the small numbers on the curve correspond to the heart beat per minute showing the direct correlation between the temperature change and the heart frequency with a shift in time between the decreasing and increasing branch of the curve due to thermal inertia of the frogs. It appears also that the following sine-like temperature oscillation is parallel to an equivalent change in the heart frequency.
Lower graph: Changes in mean frequency and standard deviation at rest of otolith No. 4. The stationary responses at constant speed of the centrifuge throughout the mission are also shown. The values are given as a percentage of those on the ground (to left of ordinate). For more details see text.

hour was well below a normal tolerance for the animal and might be partially responsible for the firing rate change at rest shown simultaneously (Fig. 8). A correction factor based on control experiments on the effect of temperature on the spike train data will take care of this possibility. (ii) From the 9th hour up to the 24th hour, owing to a leakage of oxygen, the housing pressure in the capsule went up by 11 lb in^{-2} and remained at this new level throughout the flight (Fig. 8). The possible effect of the increased pressure and increased $p(O_2)$ requires control experimentations which are now in progress.

4. Discussion and Conclusions

The first question that has to be answered is whether it is correct to apply to mammals results obtained on bull frogs. There is no doubt that as far as the single unit activity of the vestibule is concerned the difference between lower species and mammals is irrelevant [11]; in fact most of the basic work on the single unit activity of the vestibule has been performed on the ray [12]. Recent work has shown that even one of the most important secondary functions of the vestibule, namely the control of muscular postural tone is very similar in the frog and in the cat [13]. It is to be concluded therefore that the results of the OFO A experiment may be taken into account for the study of man in space.

The changes in the firing rate at rest are particularly significant for the control of muscle tone and of the vegetative nervous system, viscera, heart and circulation. A decreased vestibular spontaneous activity might provoke hypotonia and the reverse, hypertonia. Variations in muscular tone have been reported in the astronauts and explained by a lack of exercise: a vestibular component might also be involved. This altered firing rate at rest together with the lack of ability to measure head movement at a certain stage of the flight, because then the vestibular receptors do not show significant change in firing rate at rest and during maximum input, might be the cause of disorientation and of motion sickness reported in both the Russian and the American manned space flight (see Introduction). However the most important conclusion so far is that it seems that after some days the vestibular units go back to normal firing rate at rest and normal input–output function (Fig. 8), i.e. after an extended exposure to true weightlessness (input at or near zero) the vestibular organ shows a complete adaptability to the weightlessness condition itself.

We must, however, consider this adaptability as a deep modification of the functional characteristics of the organ: the disappearance of the learned pattern of response based on the constant 1 g field. This modification enables it to resume its function of measuring changes of the position of the head outside the earth gravitational field, that is with a background stimulation equal to zero. It will however not be able to measure the steady positions of the head after the movement owing to the absence of the gravitational component. As a consequence the vestibule might become unsuited to function immediately upon return to earth: it might therefore require a further period of normalization to relearn the 1 g field background pattern. This is the typical behaviour of all the sense organs. For instance, experiments have been performed, in man, reversing the optic image by a special lens in front of each eye; in this case the person becomes able to perform even complex tasks requiring the use of vision, such as skiing or driving a car, after a period of adaptation lasting 4 or 5 days. A similar period is required then for readaptation to normal vision. The fact that the vestibular organ seems to adjust to the weightlessness conditions indicates that from a practical point of view while planning long space missions (like manning orbital stations or performing flight to distant planets or establishing settlements on the celestial bodies at very low gravity, like asteroids) two choices exist: (i) Let the astronauts reach a new vestibular condition in zero gravity, thereby simplifying the plan of the spaceship or the orbital station: in this case we must remember that after the re-entry on earth a period of re-education will be indispensable: some training or drug treatment might be investigated in order to shorten the adaptability time. (ii) Try to create an environment similar to the terrestrial one by means of "artificial gravity". Of course it is obvious that this choice, besides causing new problems such as the Coriolis effects if rotation of the spaceship is used to produce an artificial gravity, is not applicable to situations of people living on celestial bodies at low gravity.

From a scientific point of view it is important not only to confirm the adaptability of peripheral organs to extremely different environmental conditions, but also to follow the time course and study the mechanism of such a process. The OFO A results show, as already reported, that adaptability of vestibular organs to weightlessness is possible: a certain number of uncertainties are still left.

(i) Though all the units studied showed an adaptability trend, the time course of such trend differs remarkably from one unit to another. The duration of the experiment, six and a half days, has not been enough to allow the observation of a complete adaptation of all units. The longest period estimated, by means of extrapolation, for adaptation of the more slowly adapting units is 10–12 days.

(ii) As it was not possible to see a sufficiently long normalized period of response after the alteration due to weightlessness, the possibility is left that the first return to normal is followed by new cycles of alteration. In this case either the cycles of alteration occur continuously and the normalization is only apparent, or normalization occurs through cycles in which the alteration gets smaller and smaller till it disappears. To solve this problem, and to test the renormalization period after re-entry on earth, a second space experiment, OFO B, is being prepared: it is anticipated that its results will be reported in one of the next COSPAR meetings.

Acknowledgments

This research was performed in part at NASA Ames Research Center during the tenure by both authors of NAS Associateships and was continued with support of a contract NAS 2211 from NASA at the II Department of Physiology of Milan University and at the CNR Space Biology Laboratory.

References

[1] V. I. Yazdovsky, I. I. Kas'yan and V. I. Kopanev, Izv. Akad. Nauk SSSR, Ser. Biol., **3**, 12 (1964).

[2] C. H. Berry, Aerospace Med. **41**, 500 (1970).

[3] B. E. GERNANDT, Exp. Neurology, **26**, 203 (1970).
[4] B. W. PETERSON, J. Neurophysiol. **33**, 750 (1970).
[5] T. GUALTIEROTTI and D. ALLTUCKER, NASA SP-115, p. 143 (1966).
[6] T. GUALTIEROTTI, Proc. VIII Int. Symp. on Space Technology and Science, Tokyo, 1969, University of Chicago Press 1971 (p. 1193).
[7] K. AKERT and B. E. GERNANDT, EEG J. **14**, 904 (1962).
[8] B. E. GERNANDT and C. A. THULIN, Am. J. Physiol. **171**, 112 (1952).
[9] B. E. GERNANDT, M. IRANYI and R. B. LIVINGSTON, Exptl. Neurol. **1**, 248 (1959).
[10] T. GUALTIEROTTI and P. BAILEY, Electroenceph. Clin. Neurophysiol. **25**, 77 (1968).
[11] E. D. ADRIAN, J. Physiol. **101**, 389 (1943).
[12] O. LOWENSTEIN and T. D. M. ROBERTS, J. Physiol. **110**, 392 (1949).
[13] F. BARALE, N. CORVAJA and O. POMPEIANO, Arch. Ital. Biol. **109**, 27 (1971).

Life Sciences and Space Research X — Akademie-Verlag, Berlin 1972

TEST OF A LIFE SUPPORT SYSTEM WITH *HIRUDO MEDICINALIS* IN A SOUNDING ROCKET

R. G. A. Lotz, P. Baum, G. H. Bowman, K.-D. Klein, R. von Lohr and L. Schrötter

Universität Frankfurt am Main, FRG

Two Nike-Tomahawk rockets each carrying two Biosondes were launched from Wallops Island, Virginia, the first on 10 December 1970 and the second on 16 December 1970. The primary objective of both flights was to test the Biosonde life support system under a near weightless environment and secondarily to subject the *Hirudo medicinalis* to the combined stresses of a rocket flight. The duration of the weightless environment was approximately 6.5 minutes. Data obtained during the flight by telemetry was used to ascertain the operation of the system and the movements of the leeches during flight. Based on the information obtained, it has been concluded that the operation of the Biosondes during the flight was similar to that observed in the laboratory. The experiment and equipment are described briefly and the flight results presented.

1. Introduction

Some years ago leeches (*Hirudo medicinalis*) were suggested for use in a long term space flight experiment [1] to determine the effect of a weightless environment on metabolism and bio-rhythms. A concept for a life support system for the experiment was proposed in 1966 [2], and during 1969 and 1970 the life support system was developed and incorporated into a unit called a Biosonde. The Biosonde was tested in the laboratory and subsequently qualified through special tests for flight on board a Nike-Tomahawk rocket.

The 86 kilogram payload was a cylinder-ogive configuration 278 cm long and 23 cm in diameter. Generally, the sections of the payload can be described as follows starting at the thrust face of the Tomahawk motor: (i) recovery and Tomahawk firing section, (ii) standard Tomahawk de-spin section, (iii) telemetry and battery section, (iv) experiment 2 Biosonde, (v) ogive nose cone containing experiment 1 Biosonde, magnetometers and ballast. Altitudes of greater than 243 km were reached in both instances and both payloads were recovered by surface vessel from the Atlantic Ocean.

2. Technical Discussion

The sub-orbital flights of the Biosonde were conducted to provide evidence that the life support system developed for the *Hirudo medicinalis* experiment would operate properly in a weightless environment. The life support system recently flown consisted of an oxygen source, a passive carbon dioxide absorber, water circulating pump and animal cage all immersed in water contained inside the airtight Biosonde. Due to the fact that the *Hirudo medicinalis* normally requires food only once per year, no feeding system is required for this experiment.

The main component of the Biosonde life support system is the oxygen producing electrolytic cell (Fig. 1). The rate of oxygen production is dependent on the amount of electrical current supplied to the electrolytic cell. A specially designed differential pressure sensor (Fig. 1) was used to control the current. The sensor

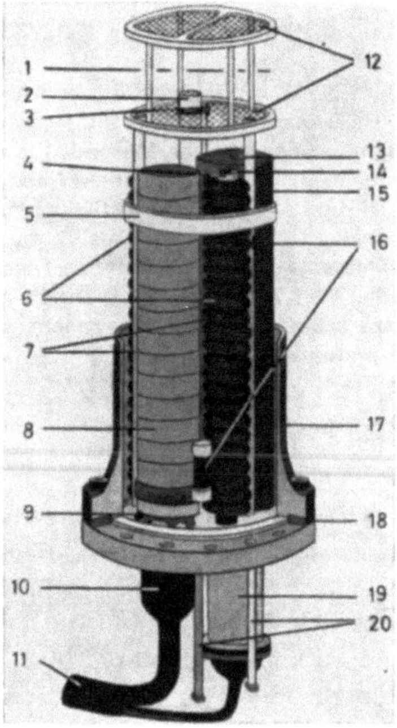

Fig. 1. Drawing of Biosonde showing integral parts: 1, animal chamber; 2, stand pipe (circulating water); 3, lamp; 4, tube; 5, metal ring; 6, CO_2 absorption; 7, stabilization rod; 8, electrolytic cell; 9, flow sensor; 10, connector; 11, cable; 12, motility sensors; 13, pressure sensor; 14, adsorption filter; 15, reference chamber; 16, temperature sensors; 17, stainless steel housing; 18, O-ring; 19, motor for pump; 20, filling tubes.

detected the difference in pressure between the Biosonde interior and the sealed reference cell (Fig. 1) built into the Biosonde. Thus, as oxygen was consumed and the carbon dioxide given off absorbed, a change in differential pressure could be detected. An appropriate amount of electric current was supplied to the electrolytic cell to replenish the oxygen supply. The oxygen produced at the electrode diffused through the silicone membrane covering the electrolytic cell and was absorbed by the water which was continuously circulated within the Biosonde. The electronic circuits were housed in a separate assembly and connected to the Biosonde by cable through a hermetically sealed connector in the base of the Biosonde.

The Nike-Tomahawk payload included two Biosondes mounted in tandem so that the longitudinal centre line of the Biosondes coincided with that of the rocket. Each Biosonde was attached to a special damping ring which decreased vibration input to the Biosonde by a factor of 4 to 6. Insulating foam was used to shield the Biosondes from anticipated temperature rise inside the nose cone during launch. The payload, mated to a "standard" yo-yo de-spin unit, was de-spun at $T_0 + 60$ seconds. Payload spin rates were measured with magnetometers mounted in the nose of the payload. A recovery system was included as a part of the payload so that a direct observation of the animals' condition could be made after the flight.

3. Flight Results and Discussion

The flight results cited here were obtained exclusively from the first Biosonde Nike-Tomahawk flight on 10 December 1970. On the second flight, telemetry drop-out occurred at $T_0 + 4$ to $T_0 + 13$ seconds and again after $T_0 + 24$ seconds. The resulting data were considered insufficient for inclusion in the data analysis.

The following parameters were monitored on board the rocket during flight and sent by telemetry to the ground receivers for recording on magnetic tape.

A. Continuously recorded data

Accelerometer (longitudinal axis $+15\,g$
$-5\,g$)

Magnetometer (X axis and Z axis)

Electrolytic cell current
Experiment 1
Experiment 2

Leech motility
Experiment 1
Experiment 2

B. Commutated data (5 Hz)

Temperatures
Two inside Experiment 1 Biosonde
One in reference cell inside Experiment 1 Biosonde
Two inside Experiment 2 Biosonde
One in reference cell inside Experiment 2 Biosonde
Two immediately outside Experiment 1 Biosonde
Two immediately outside Experiment 2 Biosonde

Motor current (water pump)
Experiment 1
Experiment 2

After the flight the continuously recorded data were digitized and stored on magnetic tape at 100 points per second for each parameter. The commutated data were digitized and then de-commutated and stored on magnetic tape at 5 points per second for each parameter. The temperatures, motor current and electrolytic cell current data were smoothed whereas the accelerometer, magnetometer and motility data were not smoothed.

The primary environmental parameter of interest is the acceleration level during the flight. In Fig. 2 the longitudinal and rotational acceleration levels are shown separately for the first 60 seconds of flight, i.e. up to de-spin. The

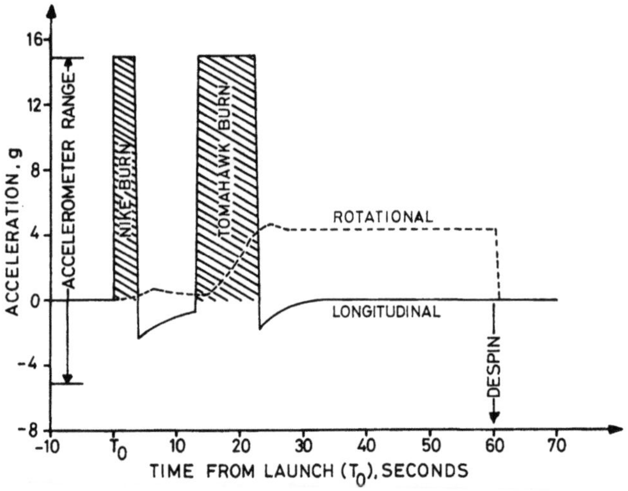

Fig. 2. Acceleration of payload for first 60 seconds of flight of Biosonde, 10 Dec. 1970: longitudinal, direct measurements; rotational, calculated using magnetometer data.

range of the accelerometer was chosen to indicate when "zero" g level was attained and, as indicated in Fig. 2, that occurred shortly after the second stage burn-out. De-spin was accomplished at 60 seconds after lift off. The rotational acceleration due to the residual spin is indicated in Fig. 3. The end of the near weightless environment was at 6.5 minutes after de-spin, at which time signals

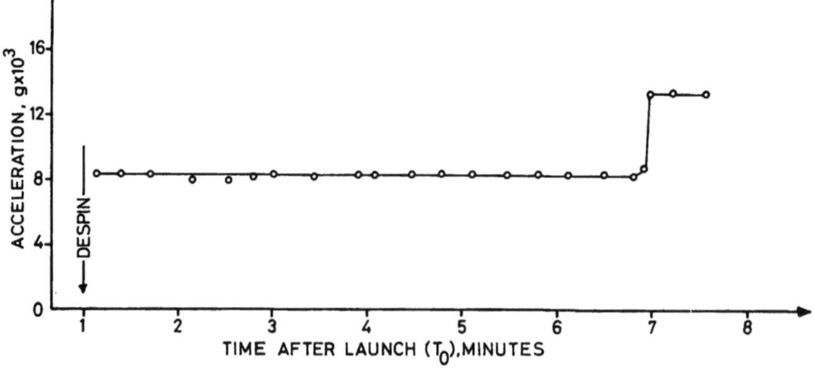

Fig. 3. Rotational acceleration of payload after de-spin. Biosonde 10 Dec. 1970.

from the second magnetometer departed from a constant value thus indicating the start of the expected flat spin.

The other parameter having direct influence on the electrolytic cell operation was temperature. In Figs. 4 and 5 the temperatures outside the Biosondes are given. It can be seen that the temperatures continued to increase throughout the low *g* portion of the flight. Based on these data there is little question but

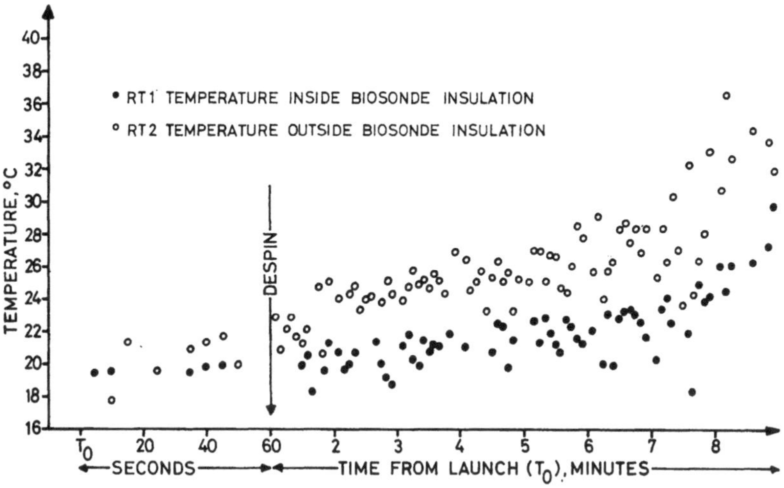

Fig. 4. Temperature of surroundings. Experiment 1 Biosonde 10 Dec. 1970.

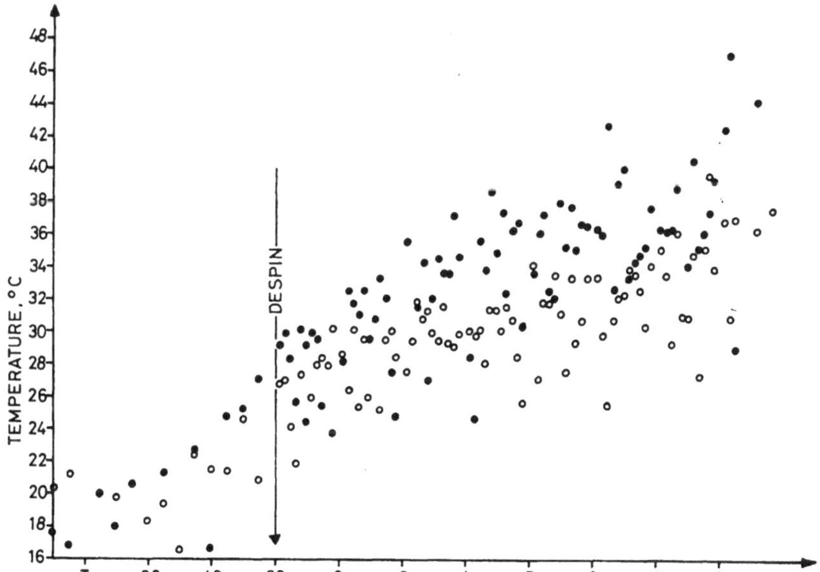

Fig. 5. Temperature of surroundings. Experiment 2 Biosonde 10 Dec. 1970.
Key as in Fig. 4

that the rocket structure absorbed a sufficient amount of heat in passing through the atmosphere to increase the temperature of the region around the Biosonde throughout the flight. Despite that increase, it is evident from Figs. 6 and 7, that the water temperature inside the Biosondes increased only a small amount during the time of interest.

The primary data, i.e. those concerning the operation of the electrolytic cell, are presented in Fig. 8. It can be seen that the current gradually increased and both cells exhibited essentially the same characteristic. The difference in slope of the two current curves is not considered significant since the sensitivities

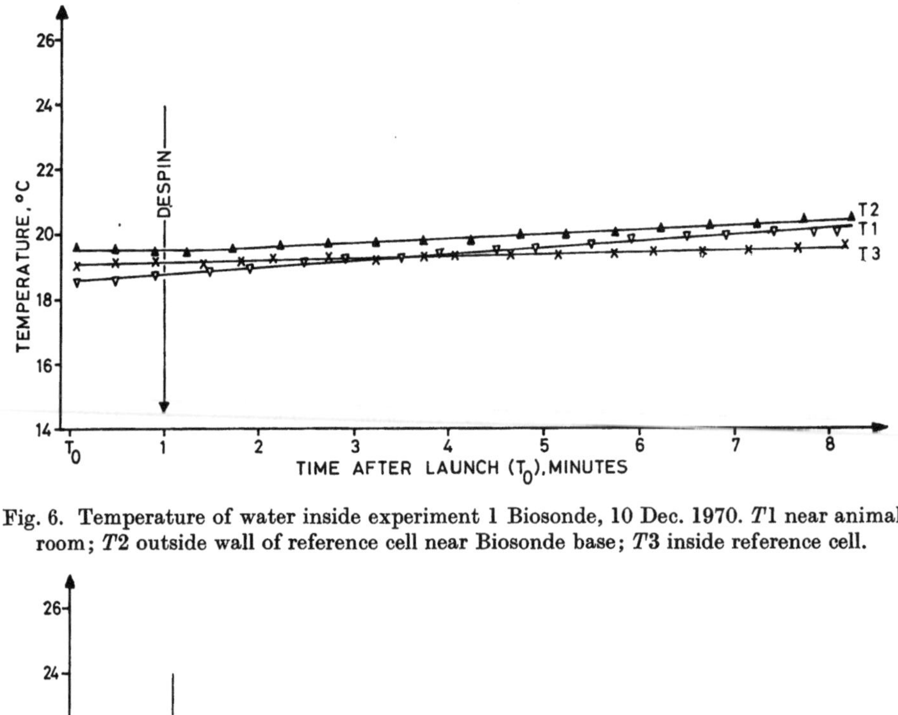

Fig. 6. Temperature of water inside experiment 1 Biosonde, 10 Dec. 1970. *T*1 near animal room; *T*2 outside wall of reference cell near Biosonde base; *T*3 inside reference cell.

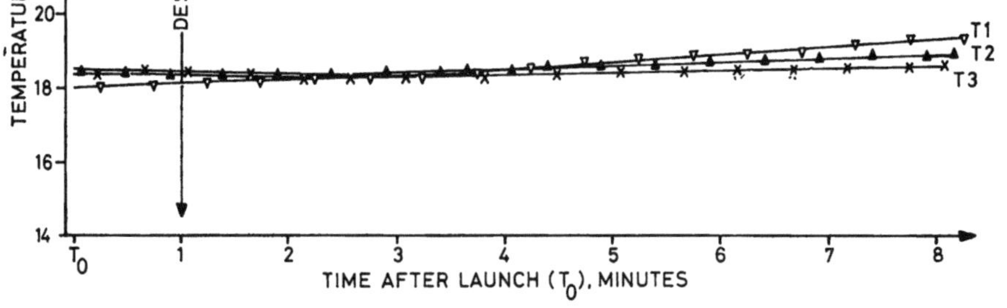

Fig. 7. Temperature of water inside experiment 2 Biosonde, 10 Dec. 1970. T1 near animal room; *T*2 outside wall of reference cell near Biosonde base; *T*3 inside reference cell.

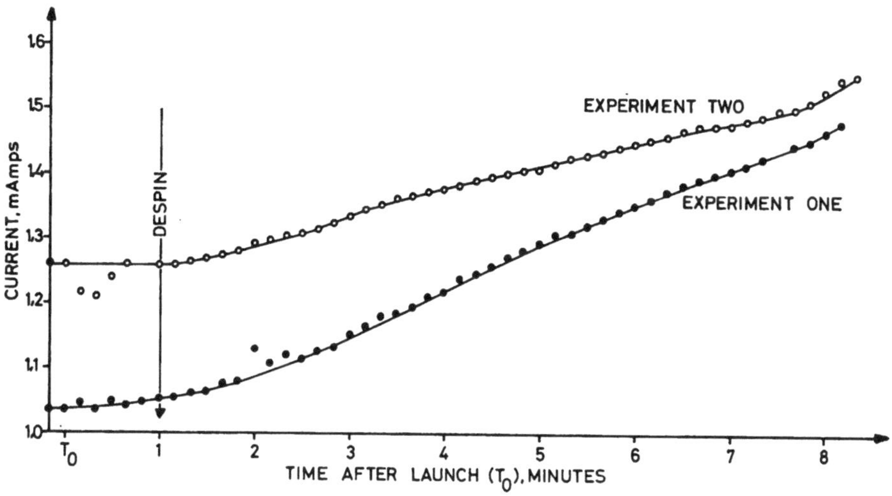

Fig. 8. Electrical current supplied to electrolytic cells. Biosonde 10 Dec. 1970.

of the pressure sensors were different. An increase in Biosonde water temperature relative to the reference cell causes a decrease in the electrolytic cell current. Similarly, if the electrodes became insulated by the oxygen produced, the current would have been reduced also. The continual increase in current implies a continual decrease in pressure which at this time can not be fully explained.

The motility data or movement of the animals within the animal rooms are shown in Figs. 9 and 10 and are included here to indicate that the overall system was compatible with the animals. It is interesting to note that at $T_0 + 7$ minutes 20 seconds both animals started to move, indicating that they may have been more sensitive to certain types of motion than the instruments used on board.

Fig. 11 shows the water pump motor current for the time of interest. The decrease which occurred about the time of maximum rotational acceleration was apparently a result of loss of magnetic coupling between the driven water pump and the motor, a condition that was later duplicated in the laboratory.

Turning now to the secondary objective, the behaviour of the *Hirudo medicinalis* during flight, it should be explained that leeches have been subjected to hard environments also [3, 4]. Specifically the experimental animal has endured acceleration levels of over 90 g, sinusoidal vibration levels greater than 20 g, random vibration levels greater than 11 g and half-sine shock levels of 60 g without any observable damage. The rocket flights should indicate, from the practical standpoint, how much the behaviour of the animal and its physiological functions will be influenced by the synergistic effects of all stress factors. Naturally, the short duration of the rocket flight does not allow any conclusions to be reached concerning long term flight in a weightless environment. On the other hand, this was the first opportunity to obtain data on this animal during weightlessness.

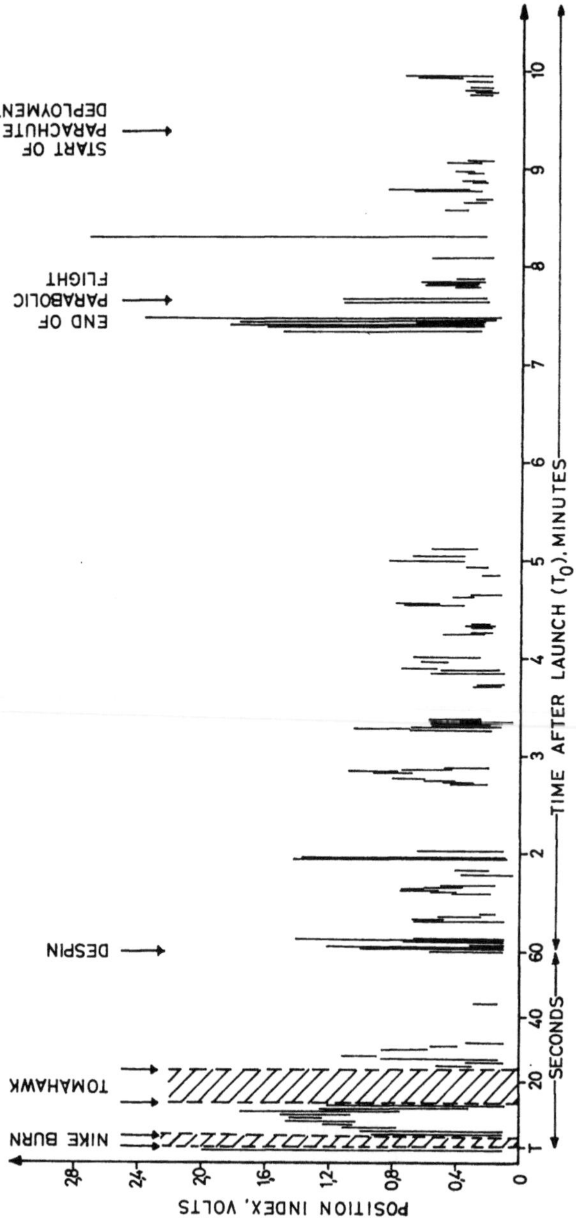

Fig. 9. Movement of leech in experiment 1 Biosonde during flight, 10 Dec. 1970.

First let us consider the motility or movement of the animals during the experiment. The animal chamber has approximate diameter and height 65 mm and 28 mm respectively. It is covered above and below by a metal net. These nets serve also as sensors for the motility measurements. The water is pumped through a tube to the animal room and flows over the whole area of the animal room, then through the nets and over the regeneration elements (electrolytic

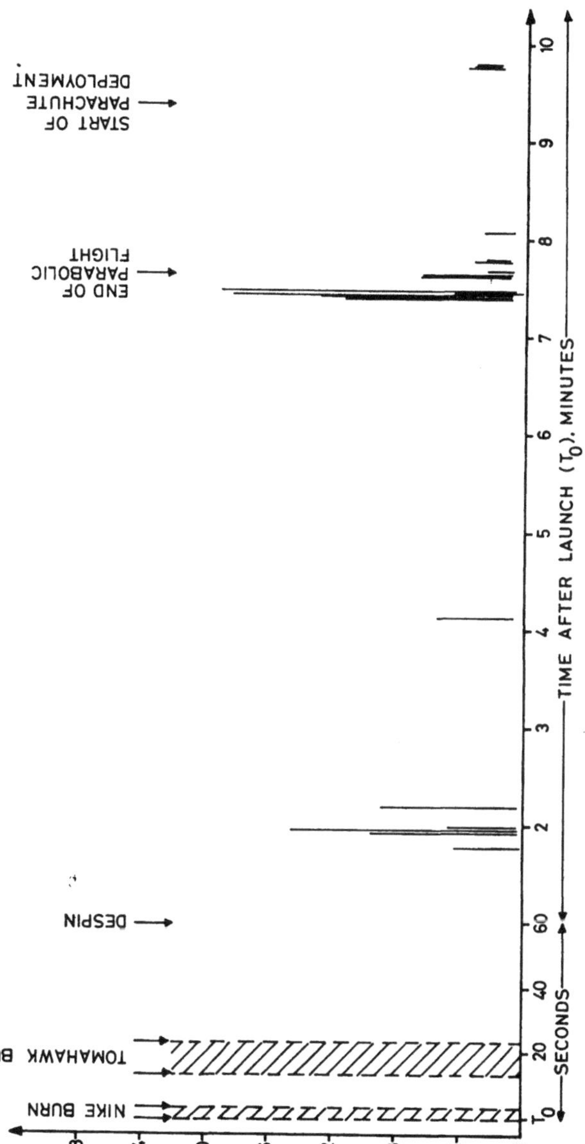

Fig. 10. Movement of leech in experiment 2 Biosonde during flight, 10 Dec. 1970.

cell, carbon dioxide absorber and active carbon filter). The animal in experiment 1 Biosonde had already before launch demonstrated a higher activity than the animal in experiment 2 Biosonde. From T_0 to $T_0 + 5$ minutes the animal in experiment 1 showed a more or less distinct motility which was interrupted only during the firing times of both rocket stages (see Fig. 9). Also during the 10 seconds between firings there was some activity. The

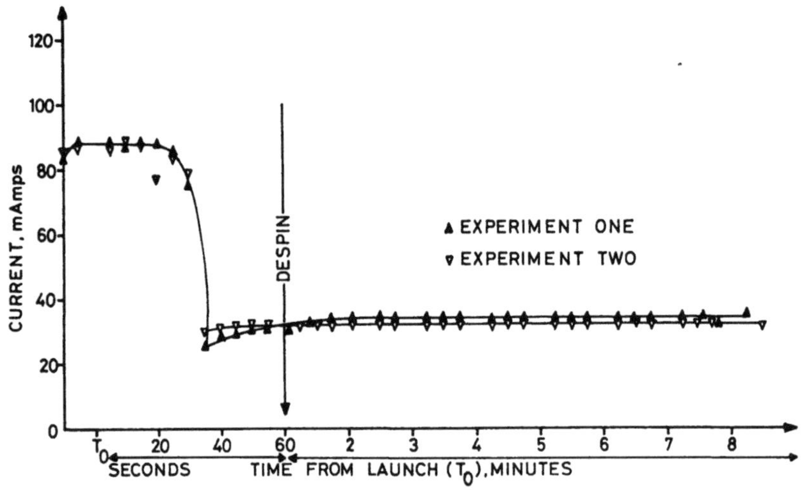

Fig. 11. Electric current drawn by water pump motors during flight. Biosonde, 10 Dec. 1970.

interruption of motility during the firing times may have been caused by the strong linear acceleration (22 g). However, the animal did not suffer any damage as can be seen in the following motility phase, i.e. prior to and after de-spin. Without doubt, both animals perceived the separation of the payload from the second stage as can be recognized by the beginning of motility at the same time after a motility pause at $T_0 + 5$ minutes. Also the opening of the parachute brought out a new activity. The less active animal in experiment 2 Biosonde (see Fig. 10) showed no movements prior to or during the launch phase. The movements began approximately one minute after de-spin and were only very weak during the whole weightless phase (6.5 minutes).

It is recognized that due to the short duration of the experiment the values of oxygen consumption must include certain errors. On the other hand, we believe that we can assume a correlation between oxygen supply and motility. Before the flight, the Biosondes were prepared in such a way that the water was not saturated with oxygen. This was done to avoid the condition of oxygen diffusing into the water too slowly due to a low partial pressure gradient. In each case the saturation of oxygen in the water was high enough for there to be no physiological stresses imposed upon the animal. In Fig. 8 the two current curves of the electrolytic cells are shown. Both curves increase during the whole time of the experiment. It can be seen that the slope of the experiment 1 Biosonde current curve is the greater, which may be interpreted as a greater depletion of oxygen from the water resulting from a greater oxygen consumption by the more active animal. This gives us at the same time an idea of the sensitivity of the pressure sensor.

The water temperature inside the Biosondes rose approximately 1 °C in the animal room and approximately 0.2 °C in the reference chamber (Figs. 6 and 7) during the whole experiment. There was no possibility that a homogeneous temperature field would be attained inside the Biosondes during the short period of the experiment (8 minutes); because of this temperature gradient, the oxygen consumption is naturally lower than is indicated by the rising angle of the electrolytic current curves, but the relation: greater activity–steeper rise will still be preserved.

In conclusion, the results of the rocket flight indicate that the Biosonde life support system operates in a near weightless environment and therefore could also be expected to operate properly in the weightless environment of space. The animals survived the changes occurring and showed the behaviour expected, especially in motility during the flight. During the weightless phase, the animals took up oxygen out of water. After the recovery, there was no damage to the animals that could be seen. Also, at this time, six months after the flight, they show no signs of delayed damage.

References

[1] R. G. A. Lotz, Life Sciences VI, 49 (1968).
[2] R. G. A. Lotz, Arbeitstagung über Extraterrestrische Biophysik und Biologie, Frankfurt 1966, Universität Frankfurt (p. 37).
[3] R. G. A. Lotz, M. Fuchs and P. Moyat, Space Life Science 2, 40 (1970).
[4] R. G. A. Lotz and G. H. Bowman, Space Life Science 2, 45 (1970).

Effects of Chronic Irradiation

Life Sciences and Space Research X — Akademie-Verlag, Berlin 1972

PHYSIOLOGICAL AND HEMATOLOGICAL EFFECTS OF CHRONIC IRRADIATION

Yu. G. Grigoryev, B. A. Markelov, V. I. Popov, A. A. Akhunov, T. P. Tsessarskaya, A. V. Ilyukhin, N. L. Fyodorova, T. E. Burkovskaya and A. V. Shafirkin

Institute of Medical and Biological Problems, Moscow, USSR

The information available for assessing radiation hazard involved in prolonged space missions is so far insufficient. Therefore, a three-year experiment was carried out in which 180 dogs were exposed to irradiation, simulating the dose value and rate of exposure that may occur in a real space flight of long duration. The exposure included a chronic irradiation (with dose rates of 21, 62 and 125 rads/year) and a combined irradiation during which the animals were exposed to chronic and acute irradiations with a dose of 8 or 42 rads applied three times every year, the annual total dose being 120 or 188 rads, respectively.

Insignificant hematopoietic changes, e.g. the ratio of red to white blood cells, their decrease in the peripheral blood, were found. Distinct changes in the reproductive function were noted. The general condition of the animals was satisfactory. Most animals endured well an additional physical load and retained conditioned reflexes developed before irradiation. The experimental findings suggest that the effects are early symptoms of the first stage of chronic radiation damage.

Publications describing the chronic effect of low dose ionizing radiation on the living organism are rather few, though information of this kind is particularly important to assay properly the radiation hazard of long-term space missions.

The radiation effect in interplanetary flights results mainly from galactic cosmic radiation manifesting as chronic exposure and from solar flare radiation acting as acute exposure [1]. According to pertinent estimates, the total dose per year's flight from galactic radiation may amount to 50–100 rem whereas doses resulting from solar flares may vary from 5–50 rem per flare [2]. Bearing this in mind, an experiment on chronic radiation was carried out. During the experiment 180 dogs were exposed to irradiation which simulated the dosage value and rate of exposure to radiation environment that may occur in prolonged space missions. The animals were irradiated in a special device whose main component was a collimator containing a ^{60}Co gamma sources. The collimator was placed above a cage where six dogs were kept simultaneously (Fig. 1). The collimator limited the irradiation field, thus ruling out the mutual effect of neighbouring radiation installations and assuring radiation safety of the personnel moving between the cages. The installations were equipped with a remote control system to transfer the radiation sources. Dosage rates within the radiation field in three different areas were 25, 75 and 150 r per year.

In addition to chronic irradiation simulating the galactic cosmic radiation effect, the animals were also exposed to acute irradiation simulating solar flare proton radiation. This exposure was to simulate dosage and time effects

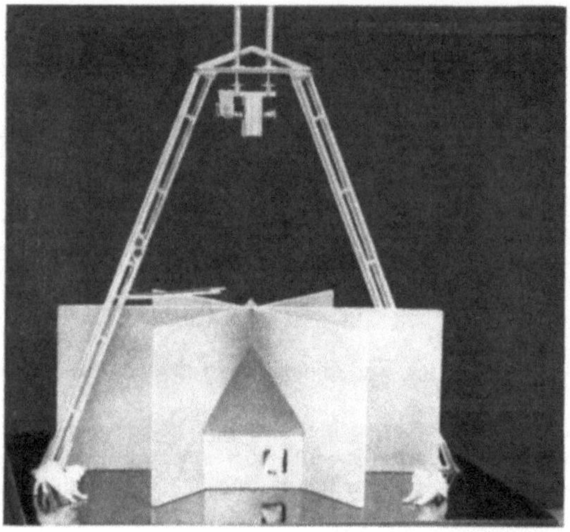

Fig. 1. Open-air cages for dogs.

of powerful solar flares, for instance the flares of 23 February 1956, 10 May 1959 and 13 February 1960. Such flares may generate doses of 5–50 rem behind a shielding of 50 g cm^{-2} and 40–50 rem behind a shielding of 10 g cm^{-2}. Having in view the possibility of three hazardous flares occurring every year, the experiment included three acute exposures a year.

When performing radiobiological experiments on large laboratory animals, it is very important to study depth-dose distribution within the body. With this purpose measurements were made on a bone–paraffin model of a dog (Fig. 2) and mid-tissue doses absorbed throughout a year were estimated. During chronic exposure mid-tissue doses were 21, 62 and 125 rads/year while during acute exposure they amounted to 8 and 42 rads/year. The irradiation procedure and the doses absorbed are given in Table 1.

The radiobiological effect of chronic irradiation was assayed on the basis of clinical, hematological, biochemical, physiological, and cytogenetic data. The present paper contains the results of a three-year experiment with chronic irradiation of dogs. The clinical state of the animals as evaluated by data obtained from regular examinations, body temperature, pulse rate, respiration rate and body weight measurements, was satisfactory throughout the whole experiment. Blood biochemical analyses (determination of the content of

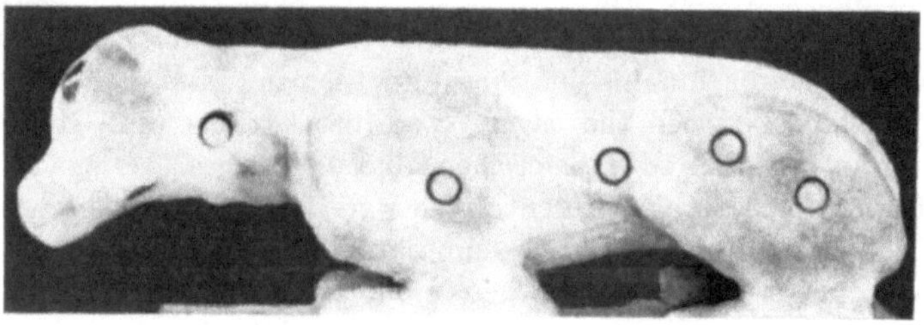

Fig. 2. Model of the dog.

Table 1
Irradiation procedure and absorbed dosage in the experiment on chronic radiation

Group No.	No. of dogs	Absorbed mid-tissue dose (rads)			
		Chronic exposure per year	Acute exposure	Total dose per year	Total dose per 3 years
1	30	21	—	21	63
2	30	62	—	62	186
3	30	125	—	125	375
4	30	62	42, 8, 8	120	360
5	30	62	42, 42, 42	188	564
6	30	Control			

protein and protein fractions, sugar, cholesterol, beta-lipoproteins, non-esterified fatty acids) showed insignificant disturbances. However, all irradiated animals displayed a lack of seasonal variation in the amount of total cholesterol. No cardiovascular disorders were observed in the test animals (V. Korchemkin). Nevertheless, a study of responses to epinephrine and acute irradiation with a dose of 42 rads revealed a decline in reactivity and a delay in compensatory reactions in the dogs of the fifth group (188 rads/year). Disease incidence among the irradiated animals was low, being comparable with that in the controls. Hematopoietic studies demonstrated certain changes in the leuco- and erythropoiesis. The changes were most clearly seen in the animals of the fourth and fifth groups (total dosages of 120 and 188 rads/year).

The changes in the leucoblastic system were apparently of the inhibitory type. As compared with the reference level, the leucocyte count in the peripheral blood of these dogs decreased to 86% by the sixth month of exposure

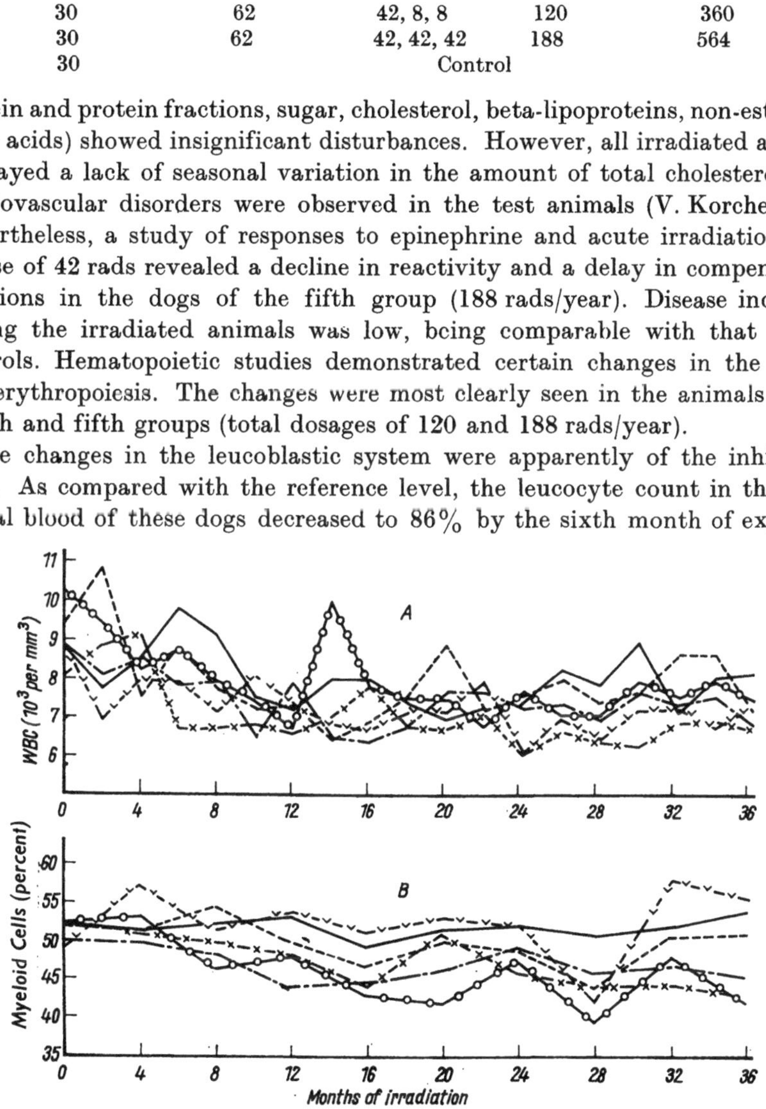

Fig. 3. Variations of the count of (A) leucocytes and (B) myeloid cells in irradiated dogs. The designations of the groups of animals is as follows:
— — — 1 (21 rad/yr); —.— 2 (62 rad/yr); —V— 3 (125 rad/yr); —O— 4 (120 rad/yr); —×— 5 (188 rad/yr).

(Fig. 3A). This decrease occurred due to a decline of both neutrophil granulocytes and lymphocytes. During this period (6 months) the bone marrow reserve of mature granulocytes also decreased to 20–60% in comparison with the control. It should be noted that throughout the three years of exposure the average level of leucocytes, neutrophils and eosinophils in the peripheral blood remained within the physiological limits, though approaching the lower value. By the end of the second year of exposure the decrease of the peripheral leucocyte count was followed by a reduction in the circulating and capillary granulocyte pools and a decrease in the half period of granulocyte circulation in the peripheral blood (to 76% compared with the control).

During irradiation the amount of myeloid elements in bone marrow decreased gradually (Fig. 3B). The difference between the experiment and the control was significant from the twelfth month ($P < 0.01$–$P < 0.001$). The decrease was evidently associated with an inhibition of proliferative processes of leucoblasts. In the sixth month the rate of DNA synthesis decreased (to 78% of the initial level) and in the twelfth month the mitotic index of myeloid elements declined significantly ($P < 0.01$). During the three-year experiment most parameters characteristic of leucopoiesis were stabilized at the level of the second year. By the end of the experiment the mitotic activity of the myeloid elements had increased appreciably (to 150–200%). Nevertheless, peripheral counts of leucocytes, neutrophils, eosinophils, granulocytic pools, bone marrow content of leucoblasts and production of granulocytes in irradiated animals were significantly lower than in controls. Therefore, despite the fact that no further decline in the leucopoietic function of bone marrow occurred, complete compensation of leucopoietic inhibition was not achieved.

Erythropoietic changes differed from those peculiar to leucopoiesis. The peripheral count of erythrocytes (Fig. 4A) and reticulocytes differed insig-

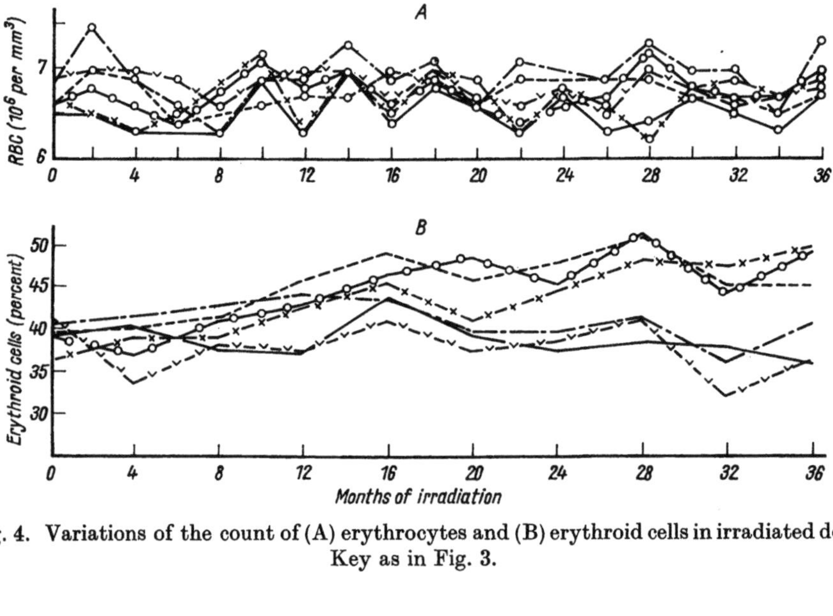

Fig. 4. Variations of the count of (A) erythrocytes and (B) erythroid cells in irradiated dogs. Key as in Fig. 3.

nificantly in comparison with the reference level, remaining within the physiological limits. However, the number of erythroid elements in bone marrow increased gradually (Fig. 4B). The difference between the experimental and control values was substantial from the twelfth month till the end of the experiment. During the third year the erythroid value was beyond the upper normal limits. Cytological study of bone marrow gave evidence that an increase in erythroid cells was due to a significant elevation of the amount of polychromatophil erythroblasts (beginning from the 16th month up to 130% compared with the control). At the same time the number of cells of early generation (proerythroblasts and basophil erythroblasts) declined slightly. By the end of the second year the fifth group dogs showed an appreciable (to 80% compared with the control) shortening of the half-period of erythrocyte circulation. This did not however bring about an anaemic syndrome. The process of accelerated death of erythrocytes was compensated for by an increase in bone marrow production of erythroid elements and mature erythrocytes (to 130% of the control). The numbers of erythrocytes entering the blood from the bone marrow during 24 hours per kg of body weight were $14.1 \pm 1.0 \times 10^9$ and $10.8 \pm 0.6 \times 10^9$ in experimental and control animals respectively. By the end of the third year the time of erythrocyte circulation and the bone marrow production returned to normal (L. Semashko et al.).

Thus, responses of myeloid and erythroid hematopoiesis to radiation effect were different. Disturbances in the erythroid system formed distinct compensatory reactions to maintain the erythrocyte count at the level required for the basic respiratory function of the blood. Erythropoietic acceleration provided that the erythrocyte count in the blood was at a high and stable level.

The lack of compensatory reactions of the leucoblast system was responsible for its reduced function. Theoretically, this may result in the limitation and disturbance of the adaptive capabilities of the system. In fact, statistical treatment of numerous hematological data demonstrated that the irradiated animals showed a lowered capacity to respond to the environmental factors which are responsible for the physiological changeability of the leucocyte picture in intact animals. The animals of groups 4 and 5 displayed smaller variations in average and individual values of the leucocyte count as well as lack of seasonal variations in the quantitative ratios between lymphocytes and neutrophils as compared with the controls. No overt irradiation-induced disturbances were detected in the thrombopoietic system. The number of bone marrow cells with chromosome aberrations was not more than 3% in all exposed animals (T. Zukhbaya). The duration of the bone marrow cell interphase remained unaltered in comparison with the control.

An investigation of reflex activity of dogs exposed to chronic irradiation with a dose of 125 rads/year showed a high radiosensitivity of their nervous system (E. Sizan). An accumulation of the total dose of 9 rads brought about an increase in the latent period of the chain motor reaction and in the number of errors. This change was observed to a greater or lesser extent during two years. By the end of the third year the number of correct responses had increased. In the 31st month (total dose of about 320 rads) the dogs were exposed to accelerations of 4 and 6 g in the centrifuge. The response of the irradiated animals was opposite to that of controls: the latent period of the early conditioned-reflex movement shortened and the number of erroneous reactions decreased in the experimental dogs and increased in the controls.

An examination of the vestibular analyzer gave evidence that its sensitivity increased and its reactivity decreased only in the dogs of the fifth group (P. Kumets).

Distinct shifts of the spermatogenic function were found in the irradiated animals. Most pronounced changes occurred in the dogs of groups 4 and 5. The amount of spermatozoa in the ejaculate was reduced at a dose of 80–100 rads (Fig. 5A). A similar value of the threshold dose was established by Casarett [3]. In addition to a reduction in the number of spermatozoa, their motility also decreased and the number of atypical forms increased (Fig. 5B, C). The number of atypical spermatozoa with primary changes also increased. This was indicative of pathological developments in the testes. Histomorphological investigations confirmed the occurrence of testicular damage. They involved depletion of the testicular tubules and hardening and collagenization of the

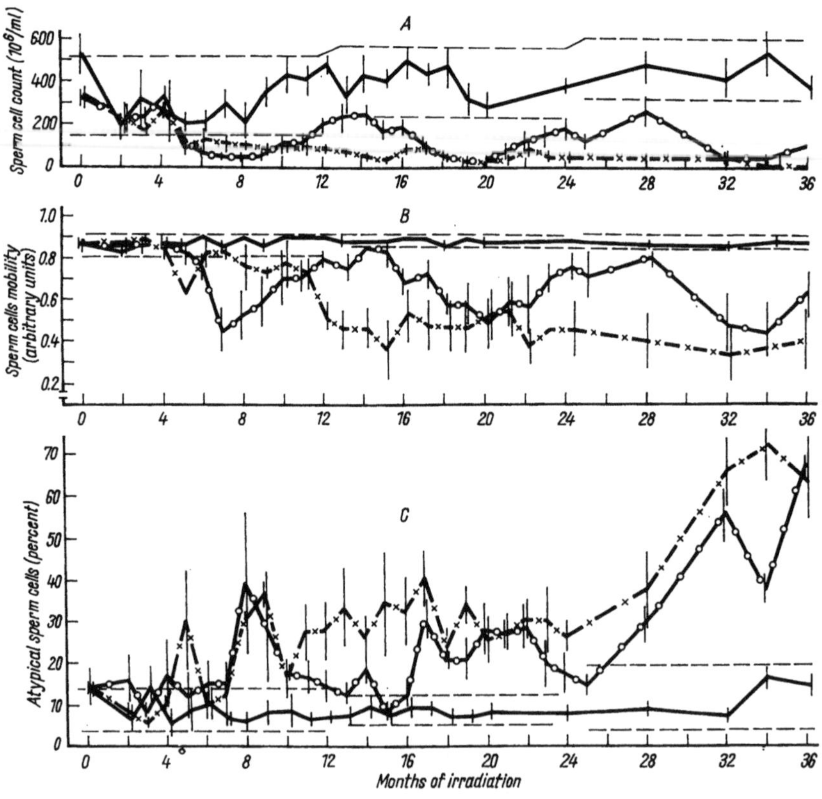

Fig. 5. Spermatogenic parameters in irradiated dogs (designations as in Fig. 3): A, the number of spermatozoa in the ejaculate; B, motility of spermatozoa; C, number of atypical forms in the ejaculate.

connective tissue of their walls. A good number of lipid drops were found in both Leydig's and Sertoli's cells. In the third year of the experiment the disturbances were most distinct. Males of the fifth group displayed sterility (total dose of 564 rads per 3 years). The sexual cycle of females exhibited no variations under the influence of the doses applied. The offspring delivered by the irradiated parents during the first two years were normal.

The experiment included a functional test of an additional physical load. After 2.5 to 3 years of exposure the load proved to be too heavy for some animals (temperature reactions and blood cell composition increased or distorted.)

Acute exposures to a dose of 42 rads performed three times every year in combination with chronic irradiations at a dose rate of 62 rads per year induced certain changes. Studies of the physiological functions revealed different responses to acute exposures that depended on the dosage accumulated by the time of the irradiation. The reactions investigated at the cell level (chromosome aberrations, DNA synthesis, mitotic activity, duration of interphase periods) showed no dependence of this type. However, as the total dose accumulated, the leucocyte reaction to repeated irradiations decreased (Fig. 6). Mathematical estimates indicated that at that period the accumulated dose was 400 rads (N. Popova, N. Appak).

Thus, the data on physiological and hematological effects of a three-year chronic irradiation of large laboratory animals (180 dogs) have been obtained. The experiment has been conducted to involve dosage rates ranging from 21 to 188 rads/year; by the end of the experiment total doses of 62–564 rads

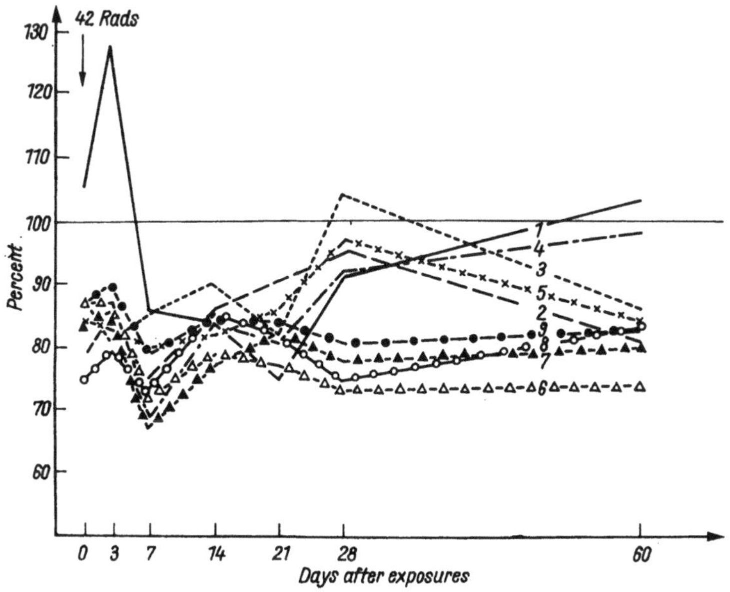

Fig. 6. Changes in the leucocyte count of the peripheral blood of dogs exposed to acute irradiations with a dose of 42 rads (the numbers 1–9 are the numbers of repeated acute exposures).

have been accumulated. Early symptoms of radiation injury and decreased capabilities for compensating disturbed functions have been revealed.

References

[1] V. E. Doodkin et al., in: Biological Effect of High Energy Protons, Atomizdat, Moscow 1967 (p. 8).
[2] Yu. G. Grigoryev et al., Kosm. Biolog. i Med., 2, 3 (1967).
[3] G. Casarett, in: Effect of Ionizing Radiation on the Reproduction System, New York, Paris 1964.

Life Sciences and Space Research X — Akademie-Verlag, Berlin 1972

THE EFFECTS OF DOSE PROTRACTION ON HEMATOPOIESIS IN THE PRIMATE AND DOG

J. F. Spalding, L. M. Holland and J. R. Prine

Biomedical Research Group, Los Alamos Scientific Laboratory,
University of California, Los Alamos, New Mexico, USA

The modifying effects of dose protraction by fractionation or continuous low dose-rate exposure are not well known in the primate. Comparative studies between the two mammalian species dog and monkey with widely differing acute LD_{50} values have not been made. A program to study dose-rate effects on injury and recovery of the bone marrow in dogs and monkeys under simultaneous and similar exposure conditions is in progress. The equivalent residual dose (ERD) assumptions of 10% irreparable injury and a 28-day recovery half-time for the reparable portion (90%) of radiation were studied in dogs (beagles) and monkeys (*Macaca arctoides*) using variable recovery times and appropriate fractionated exposures to attain but not exceed theoretical effective residual body burdens of 200 rads over a 1-year period. The ERD assumptions were tolerated by the dogs but resulted in death to 7 of 8 monkeys tested. The same ERD assumptions were tested with monkeys (*Macaca mulatta*) using a fixed recovery time span and an upper ERD limit of 100 rads. Monkeys in this study have tolerated 1900 rads of gamma-ray exposure (approximately 3 times the acute lethal dose) with minimal suppression of the hematopoietic system.

1. Introduction

The equivalent residual dose (ERD) concept for predicting the biological and medical consequences of exposure to ionizing radiation has been used by the National Committee on Radiation Protection and Measurements [1]. The ERD assumptions for human exposure to ionizing radiations are as follows: (i) 10% of the dose-related injury is irreparable; (ii) the remaining reparable injury (90%) heals with a repair half-time of one month, and all possible recovery has occurred after three months; (iii) recovery is continuous during protracted exposure; and (iv) radiation injury is proportional in magnitude to the brief dose that causes the injury. If these assumptions are reasonably accurate, both acute (brief) and protracted exposures so regulated that the ERD never exceeds 200 rads should be tolerated by man for up to a year or more without great effect. The ERD concept of radiation injury and repair was adopted to provide a more reliable method for predicting radiation injury than does the use of total accumulated dose without regard to dose rate. Because no research experience was available to test the validity of the ERD concept, it was not recommended for predicting the outcome of exposures longer than one year. The need for experimental tests of the ERD concept and dose-rate protraction in mammalian species similar to man is obvious. Because the experimental species most closely resembling man has not been clearly identified, the use of two or more species (one related to man and recommended as the best "all-purpose laboratory primate" [2]) with different radiation-response characteristics [3] should yield useful data for evaluation of the ERD concept and its accompanying assumptions.

A dog–primate study to test the upper limits of the ERD assumptions stated above showed that the acute lethal gamma-irradiation dose of the monkey (*Macaca arctoides*) was increased approximately 200 rads by dose protraction (fractionation method). Seven of eight monkeys tested died from radiation-induced bone-marrow injury with theoretical ERD levels below 200 rads [4], suggesting that dose protraction by fractionation may not attenuate the effects of radiation injury so much as the ERD assumptions for man [1] would indicate. Dose protraction with monkeys (*Macaca mulatta*) and dogs (beagles) is being investigated to determine the usefulness of the ERD concept and exposure-condition limitations. Current data from these studies are presented.

2. Methods

Twelve monkeys (*Macaca mulatta*), eight females and four males approximately four years old, were subjected to an initial brief gamma-ray exposure ($\sim$40 rads/h) of 100 rads. Fifty-six days after the initial 100-rad challenge dose and at 56-day intervals thereafter, the animals are being stressed with additional 100-rad exposures. The monkeys are bled from the saphenous vein frequently before and after each exposure, and peripheral blood characteristics are observed. This is a continuing study; data through the nineteenth 100-rad exposure are presented. A species-comparison investigation is also being done to study the effects of dose protraction on hematopoiesis in dogs and monkeys. Six dogs (beagles) and six monkeys (*Macaca mulatta*) were given simultaneous whole-body gamma-ray exposures totaling 660 rads protracted over 237 h (2.78 rads/h), a total dose and dose rate calculated to bring the dog to the brink of hematopoietic death. Peripheral blood characteristics of these animals were studied before, during, and after exposure. Eighty-four days after exposure (a recovery time suggested for complete repair of radiation-induced reparable injury in man [1]), all animals of both species were exposed continuously to ^{60}Co gamma irradiation at 1 rad/h. A species comparison of the hematopoietic effects of protracted gamma-ray exposure with these two dose rates is being made. The data from the first exposure (2.78 rads/h) and through 62 days of the second protracted exposure (1 rad/h) are presented here. Dose rates for these investigations are mid-body air doses determined using 0.25 R, 1.0 R and 2.5 R high-energy, Victoreen cobalt chambers cross-calibrated with each other and with the National Bureau of Standards.

3. Results and Discussion

3.1. 100-rad Gamma-ray Fractions at 56-day Intervals

Packed cell volume (PCV) and white blood cell (WBC) counts of blood samples taken from the saphenous vein have been plotted in Fig. 1. Data from all blood samples through 363 days, including seven 100-rad gamma-ray exposures

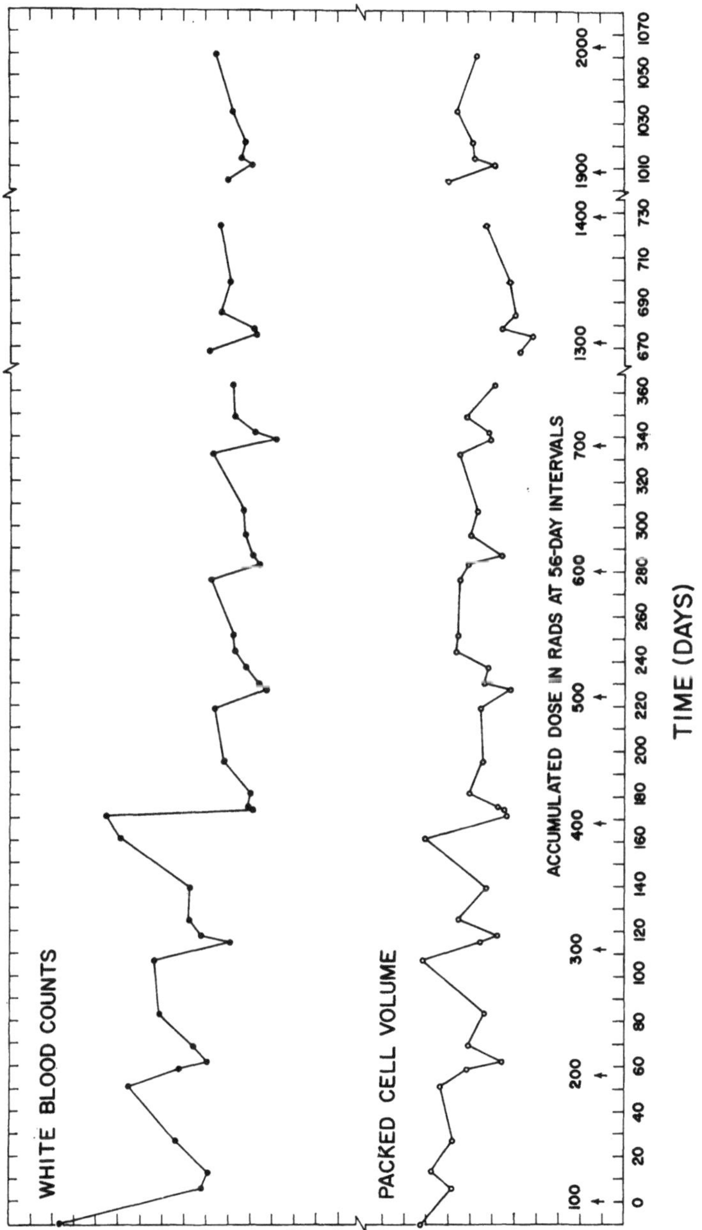

Fig. 1. Mean white blood cell (WBC) count and packed cell volume (PCV) of 12 monkeys plotted against time during a 360-day fractionated gamma-ray exposure regime that included seven 100-rad exposures given at 56-day intervals (PCV and WBC response to thirteenth and nineteenth 100-rad exposures).

are plotted. Mean PVC's were affected somewhat by each 100-rad exposure, but at no time during the first year did the mean PCV of the twelve monkeys drop below 44%. The bone-marrow response to gamma-ray injury, as reflected by PCV values, did not appear to be affected or attenuated by the accumulation of 700 rads of gamma rays given during approximately one year (Fig. 1). The more radiosensitive WBC's responded dramatically to each of the 100-rad gamma-ray exposures, as shown in Fig. 1. Following 400 rads of gamma-ray exposure, WBC counts dropped to somewhat depressed levels and have remained there. Peripheral blood elements have been observed for 1035 days, during which time the monkeys have been subjected to nineteen 100-rad exposures. Bone-marrow injury and recovery responses from the thirteenth (~2 years) and nineteenth (~3 years) 100-rad exposures, as reflected by PCV and WBC values of peripheral blood samples, are also shown in Fig. 1. The PCV was generally lower than normal toward the end of the second year (thirteenth 100-rad exposure), but at the end of the third year (nineteenth 100-rad exposure) it was at or near pre-exposure values. The WBC counts have remained at essentially the levels that were reached after the fourth 100-rad exposure (Fig. 1). The pattern of response to and recovery from 100-rad gamma-ray exposures after the nineteenth 100-rad exposure was not markedly different from that following the first 100-rad exposure.

If fractionation of gamma-ray dose is equivalent to continuous exposure to the same total dose over the same time span (fractionation of a given dose has generally been found to be more biologically damaging than continuous protraction), then 100 rads at 56-day intervals would be equivalent to approximately 1.8 rads/day. Using this rationale and the equivalent residual dose (ERD) assumptions discussed earlier (and specifically stated below), one can plot the theoretical ERD for this investigation against time.

(i) 10% of the radiation-induced injury (dose) is irreparable.

(ii) The remaining 90%, or reparable injury, heals at a rate of 2.5%/day (approximate RT_{50} of 28 days).

(iii) Recovery from an acute exposure does not begin until four days after exposure.

(iv) During protracted exposure, recovery is continuous.

(v) The ERD at t days can be obtained from the expression

$$\mathrm{ERD} = D_0\,[0.1 + 0.9\,(1.000 - 0.025)^{t-4}] + D \int_4^t [0.1 + 0.9\,(1.000 - 0.025)^{t}]\,dt,$$

where D_0 = acute dose in R received initially, D = protracted daily dose at a constant rate, R per day, t = time in days after start of initial exposure. Eq. (1) may be expressed as $\mathrm{ERD} = a(t)\,D_0 + b(t)\,D$, where a is a multiplier for a brief dose appropriate to time t and b a multiplier for a protracted dose rate appropriate to time t.

The ERD for a protracted gamma-ray exposure of 1.8 rads/day following a brief 100-rad exposure is plotted against time in Fig. 2. The PCV and WBC counts shown in Fig. 1 have been replotted in Fig. 2 with the direction of the ordinate axis reversed to express hematological response to radiation exposure as an injury increase and recovery as an injury decrease to demonstrate whether

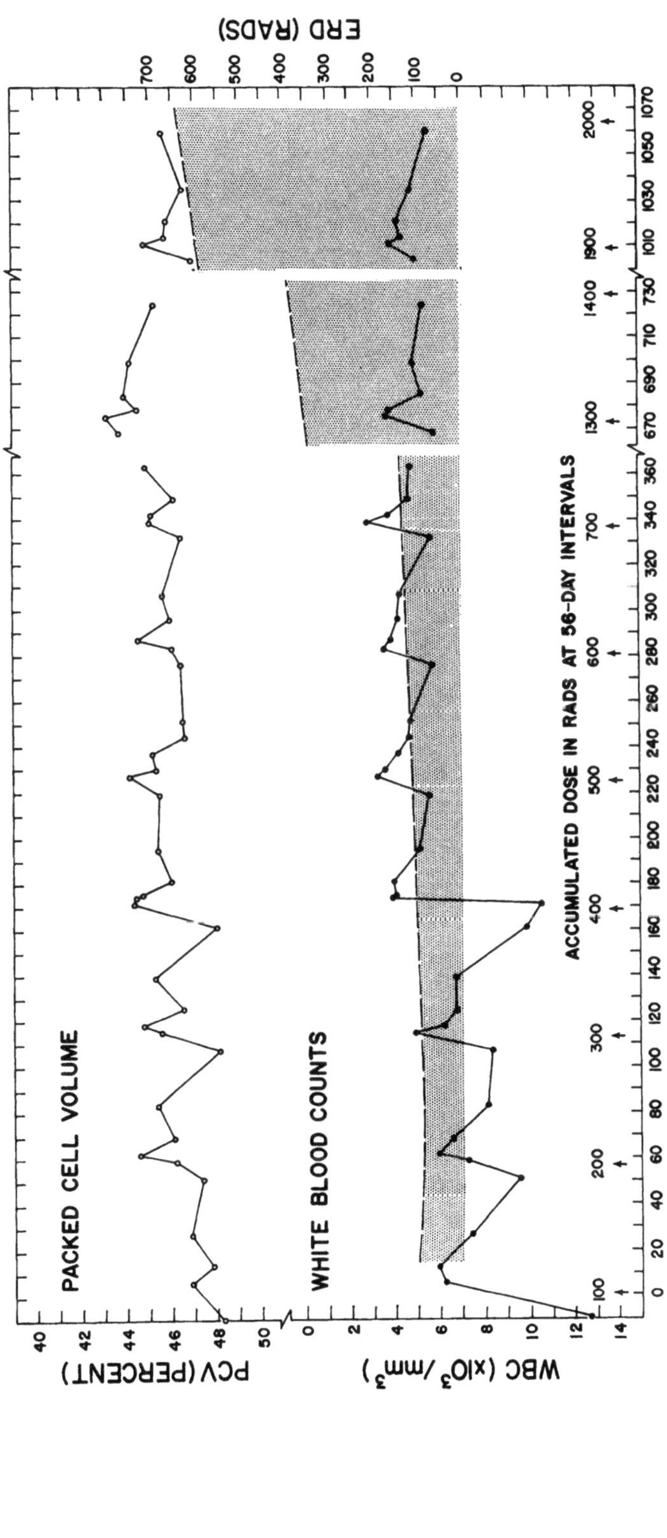

Fig. 2. Mean white blood cell (WBC) count and packed cell volume (PCV) plotted against time with the direction of the left-hand ordinate axis reversed relative to Fig. 1. Theoretical equivalent residual dose (ERD) is plotted against time using the right-hand axis for the ordinate. The ERD is plotted assuming continuous exposure at a dose rate of 1.8 rads/day.

a relationship between ERD and hematological injury exists. The theoretical ERD value drops from 100 rads on the day of the brief exposure to a minimum of 85.67 rads on day 62. From day 66 the ERD increases gradually, and on day 363 it is 132.45 rads (Fig. 2). The ERD curve is a reasonably good fit for both the WBC count and PCV through the first year of exposure. The theoretical ERD increases from 354 to 389 rads between 665 and 735 days and from 591 to 632 rads between 1000 and 1070 days (Fig. 2). No relationship between theoretical ERD and hematopoietic injury was evident at the end of the second or third year of exposure. On day 752, one animal died with symptoms similar to the acute radiation death syndrome. Two hours before death this animal had a PCV of 23.5 and a WBC count of 5900, and death was attributed to generalized subcutaneous and visceral capillary hemorrhage. The theoretical ERD at time of death was 510 rads (410 rads from the fractionation investigation in progress and 100 rads of irreparable injury acquired during an earlier study [5] in which this monkey received 1000 rads of gamma-ray exposure protracted over ten days). At this point in the study (1060 days), the ERD is 632 rads, a dose within the acute LD_{50} range for the monkey [6–12].

3.2. Species Comparison Investigation

Packed cell volume (PCV) and white blood cell (WBC) counts of peripheral blood from dogs and monkeys exposed simultaneously to 660 rads of gamma rays protracted over ten days are plotted in Fig. 3. The accumulated gamma-ray dose received during this exposure is 2–3 times the acute LD_{50} reported for the dog [13–19] and within the acute LD_{50} range of the monkey [6–12]. Assuming recovery to be continuous during the above exposure and using the ERD factors described earlier we find an ERD of 537 rads at the end of this protracted exposure (Fig. 4). This theoretical equivalent dose is still twice the acute LD_{50} reported for dogs. None of the six exposed dogs died, so the ERD factors used were apparently conservative for this radiation regime. Although no dogs or monkeys died after this exposure, three of the six dogs hemorrhaged internally, and their condition was considered critical. The PCV and WBC counts in the dogs dropped to their lowest levels (WBC $\sim$11% and PCV $\sim$70% of pre-exposure levels) about three weeks after exposure (Fig. 3).

Bone-marrow response in monkeys was quite different from that of the dog. The PCV and WBC counts were lowest about two weeks after exposure, with WBC dropping to approximately 38% and PCV to approximately 90% of pre-exposure values. At no time following this exposure did the monkeys show signs of radiation sickness. It was apparent from the peripheral blood observations that the dog's bone marrow was more radiosensitive to this dose rate (2.78 rads/h) than that of the monkey (Fig. 3), an observation consistent with the literature [3]. Peripheral blood elements (PCV and WBC) of dogs and monkeys had stabilized near pre-exposure levels within 60 days after the animals were removed from the exposure environment (Fig. 3).

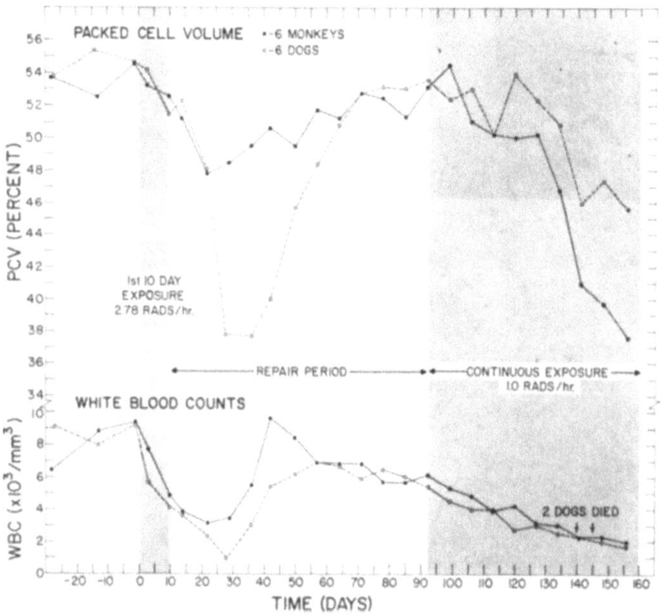

Fig. 3. Mean white blood cell (WBC) count and packed cell volume (PCV) of dogs and monkeys plotted against time before, during, and after gamma-ray exposures at two dose rates: 660 rads at 2.78 rads/h, and continuous exposure at 1.03 rads/h (investigation still in progress).

The existing data suggest that dose protraction at low dose rates increases the median lethal dose in the monkey less than in the dog. According to the literature [3], the monkey shows a small, gradual increase in LD_{50} with decreasing dose rate but fails to show a dose-rate LD_{50} inflection point at 1.20 rads/h, whereas the dose-rate LD_{50} of the dog shows an inflection point at about 6 rads/h. The ERD assumptions in [1] suggest that essentially all recovery from an acute exposure has occurred after three months. However, the theoretical ERD 84 days after 660 rads of exposure is approximately 110 rads (Fig. 4). Monkeys and dogs that had recovered to this assumed ERD were exposed simultaneously to ^{60}Co gamma rays at a dose rate of 1 rad/h to determine the comparative effect of dose protraction on median lethal dose in these species and the validity of ERD assumptions for each species. Theoretical ERD values plotted against time during the two exposures in this comparative study are shown in Fig. 4. At a dose rate of 2.78 rads/h (63.94 rads/23 h day), the ERD builds up steeply and reaches approximately 537 rads after 237 h of exposure or an accumulated dose of 660 rads. The ERD drops to approximately 110 rads after 84 days of recovery. Starting with this ERD, we administered continuous ^{60}Co gamma-ray exposure at 1.03 rads/h; the theoretical ERD for this dose rate is plotted through 62 days of exposure in Fig. 4. The PCV and WBC counts of peripheral blood from dogs and monkeys during the protracted exposure at 1 rad/h are shown in Fig. 3 through 62 days (investigation still in progress).

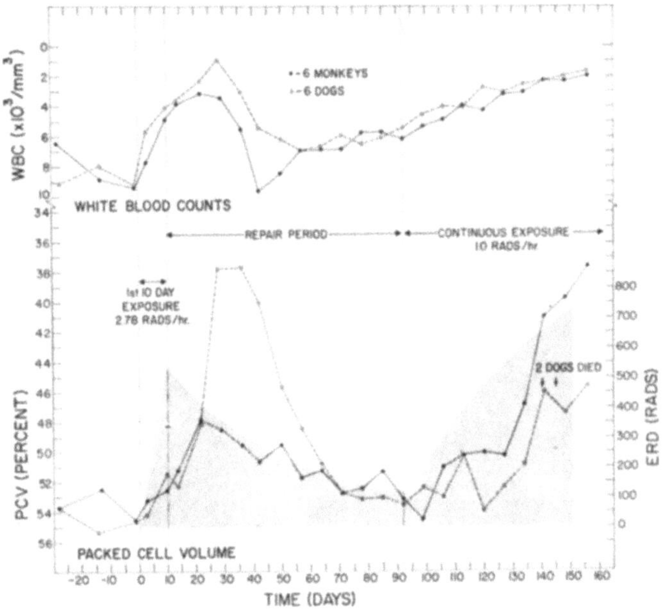

Fig. 4. Mean white blood cell (WBC) count and packed cell volume (PCV) of dogs and monkeys plotted against time with the direction of the left-hand ordinate axis reversed relative to Fig. 3. Theoretical equivalent residual dose (ERD) is plotted against time using the right-hand axis for the ordinate during and following gamma-ray exposures at two dose rates: 660 rads at 2.78 rads/h, and continuous exposure at 1.03 rads/h (investigation still in progress).

The data through 62 days of exposure show that dose protraction at 23 rads/day provides a more dramatic increase in median lethal dose for the dog than for the monkey. This observation is consistent with the literature [3]. Although this investigation is still in progress, the blood picture indicates that the mean accumulated dose (MAD) of gamma rays at death for both the monkey and dog is 1650–1850 rads. If the dose absorbed during the 14 days of exposure before death is considered an overkill dose or wasted radiation [3], the median lethal dose (MLD) for both species at this dose rate would range between 1318 and 1518 rads. Thus, dose protraction at 1 rad/h would increase the LD_{50} for the dog by a factor of approximately 4 and that for the monkey by approximately 2. The theoretical ERD for an accumulated dose range of 1318 to 1518 rads at this dose rate is 730–780 rads, which is somewhat higher than the acute LD_{50} for the monkey.

The acute LD_{50} for man is estimated to be approximately 450 rads, intermediate between that of dog and monkey, and hematopoietic recovery from protracted gamma-ray exposure totaling 2–4 times the acute lethal dose in man appears to be complete within 60 days after exposure [20], as with the dog and monkey. Thus, man might reasonably be expected to react to protracted radiation exposure in a manner intermediate between those of the dog and the monkey. The ERD curves for the dose rates and recovery times used in the species-comparison investigation are plotted in Fig. 4 with the PCV and WBC counts treated as in Fig. 2 for the same reason. Under the dose-rate and repair conditions of this comparative study, the ERD concept appears to be a reasonably good predictor of hematopoietic injury and recovery (Fig. 4).

4. Summary and Conclusions

Two investigations are in progress to study the effects of dose protraction on hematopoiesis in dogs and monkeys and to test the usefulness of the ERD concept for predicting radiation-induced injury and recovery. Data currently available from these investigations suggest that (i) the dog is more radiosensitive to acute gamma-ray exposure than is the monkey; (ii) the dog is more responsive than the monkey to dose protraction in terms of an increase in MLD; and (iii) the ERD concept is a reasonably good indicator of hematopoietic injury and recovery for the exposure conditions used in this investigation. If the upper ERD limit recommended in [1] is lowered from 200 to 100 rads, the ERD concept should be a reasonable and conservative guide for exposure of man to radiation in an emergency.

Acknowledgments

The authors are grateful to O. S. Johnson for dosimetry and technical assistance, to R. F. Archuleta for facility design and fabrication, to Norma J. Basmann for technical illustrations and services, and to Patricia M. LaBauve, Jerry E. London, R. H. Wood, and E. A. Vigil for technical assistance. We also thank Gary L. Tietjen of the Theoretical Division for statistical services in providing equivalent residual dose programs.

This work is being performed under the auspices of the US Atomic Energy Commission.

References

[1] National Committee on Radiation Protection and Measurements, Exposure to Radiation in an Emergency, Nat. Bur. Stds. Rep. No. 29, Washington, D. C., 27 October 1961.
[2] C. H. KRATOCHVIL, Ann. N. Y. Acad. Sci. **162**, 71 (1969).
[3] N. P. PAGE, in: Proc. Symp. on Dose Rate in Mammalian Radiation Biology, UT-AEC Agricultural Research Laboratory and U. S. Atomic Energy Commission Rep. No. CONF-680410, 12 July 1968 (p. 12.1).
[4] J. F. SPALDING, L. M. HOLLAND and O. S. JOHNSON, Health Phys. **17**, 11 (1969).
[5] J. F. SPALDING et al., in: Proc. Symp. on Dose Rate in Mammalian Radiation Biology, UT-AEC Agricultural Research Laboratory and U. S. Atomic Energy Commission Rep. No. CONF-680410, 12 July 1968 (p. 9.1).
[6] E. ELDRED and W. V. TROWBRIDGE, Radiology **62**, 65 (1954).
[7] M. V. HAIGH and E. PATTERSON, Brit. J. Radiol. **29**, 148 (1956).
[8] U. K. HENSCHKE and J. L. MORTON, Am. J. Roentgenol. **77**, 899 (1957).

[9] R. G. Allen et al., Radiation Res. **12**, 532 (1960).
[10] G. V. Dalrymple, I. R. Lindsay and J. T. Ghidoni, Radiation Res. **25**, 377 (1965).
[11] J. R. Eltringham, Radiation Res. **31**, 533 (1967).
[12] R. E. Stanley, L. J. Seigneur and T. A. Strike, Armed Forces Radiobiology Res. Inst. Rep. No. SP-66-23 (December 1966).
[13] C. A. Gleiser, Am. J. Vet. Res. **14**, 284 (1953).
[14] V. P. Bond et al., Radiation Res. **4**, 139 (1956).
[15] J. N. Shively, S. M. Michaelson and J. W. Howland, Radiation Res. **9**, 445 (1958).
[16] C. L. Hansen, Jr., S. M. Michaelson and J. W. Howland, Public Health Rep. No. 76, 1961 (p. 242).
[17] J. N. Shively, S. M. Michaelson and J. W. Howland, Radiation Res. **15**, 319 (1961).
[18] E. J. Ainsworth et al., Radiation Res. **26**, 32 (1965).
[19] S. M. Michaelson et al., in: Proc. Symp. on Dose Rate in Mammalian Radiation Biology, UT-AEC Agricultural Research Laboratory and U. S. Atomic Energy Commission Rep. No. CONF-680410, 12 July 1968 (p. 7.1).
[20] R. G. Martinez et al., Rev. Med. Inst. Mex., Seguro Social **3**, 14 (1964).

Life Sciences and Space Research X — Akademie-Verlag, Berlin 1972

SUMMARY OF LATENT EFFECTS IN LONG TERM SURVIVORS OF WHOLE BODY IRRADIATIONS IN PRIMATES

J. H. Kirk, H. W. Casey and J. E. Traynor

USAF School of Aerospace Medicine, Aerospace Medical Division (AFSC)
Brooks Air Force Base, Texas, USA

The USAF School of Aerospace Medicine, Radiobiology Division, Brooks Air Force Base, Texas presently is maintaining a colony of over 450 primates in which the whole body has been exposed to various types of space radiation including protons and electrons. The majority of the primates (*Macaca mulatta*) were exposed during 1965. Types of radiation involved are 2 MeV X-rays, 5 MeV–2.3 GeV protons and 1.6 MeV electrons. Low energy proton dose range up to 3000 rad (50–100 rad min^{-1}) whereas the penetrating energy doses range up to 700 rad (15–100 rad min^{-1}). Primates from a simulated solar flare exposure are also included. In late 1970, a small group of primates exposed to 108 and 85 MeV alpha particles (eye and partial body only) were added to the colony. Data are available in the following areas: (i) chronic skin changes; (ii) testicular atrophy; (iii) cataractogenesis; (iv) hematological and serum biochemical analysis; (v) incidence of tumors; (vi) causes of death; (vii) body weight variations; and (viii) summary of alpha particle experiences.

1. Introduction

For many years the USAF School of Aerospace Medicine, Brooks Air Force Base, Texas, has been participating with the National Aeronautics and Space Administration to conduct research on space radiation effects. This paper concerns the study by the Radiobiology Division to determine, what, if any, latent effects may result from exposure to space type radiations. The study, therefore, deals primarily with protons which comprise the major component of galactic, cosmic and solar radiations [1, 2]. Additional studies utilizing X-rays, electrons and alpha particles are being conducted.

In the present paper those conditions which exist in the colony at present are discussed under the following headings: present numerical colony status, latent skin effects, testicular effects, cataractogenesis, hematological and serum biochemical data, tumor occurrences, causes of death, body weight variations, and summary of alpha particle experiences.

2. Present Status

The "Chronic Colony" consists of over 400 *Macaca mulatta* animals which represent the survivors from an original group of over 1000 primates [3–7]. Maintenance and care of these animals is performed by the Veterinary Sciences

Division, USAF School of Aerospace Medicine, Brooks AFB, Texas. The primates are housed individually, for the most part, in out-of-doors chain link cages covered by a solid roof. They are fed a commercially available monkey biscuit containing isoniazid, which is supplemented with apples or bananas once weekly. Water is available at free choice. The following groups of primates, listed according to exposure are being maintained:

Skin penetrating protons (5–28 MeV)

Group	Date exposed	Numbers	Dose range (rad)
Oak Ridge I	June 1967	26 (5)*	2000–500
Oak Ridge II	March–April 1968	7 (4)	1000
Texas A & M	November 1969	18 (3)	3000–1000

* Number of control animals

Penetrating protons

Group	Date exposed	Numbers	Dose range (rad)
32 MeV	July 1964	11 (1)	560 and 280
55 MeV	April–May 1965	57 (6)	600–25
138 MeV	January 1965	25 (5)	650–210
400 MeV	March 1965	51 (9)	600–50
2.3 GeV	October 1965	48 (7)	560–50

Solar flare

90% — 10 MeV protons	Exposed April 1969	
10% — 105 MeV protons	27 (7)	1200–300 rad

Alpha particles — October 1970

Group	Energy	Numbers	Dose range (rad)
Single eye	108 MeV	11	1000–250
Single eye	85 MeV	11	1000–250
Skin only	108 MeV	12	2000–1000
Visual task	108 MeV	4	1000

Other exposure groups

Group	Date	Numbers	Dose range (rad)
2 MeV X-rays	March–April 1964	29 (1)	716–400
1.6 MeV electrons	May 1968	10	1500 and 1000
2 MeV electrons	November 1969	12 (4)	1500–900

Animals were rotated in lucite chairs or mesh wire cylinders during exposure to diffuse the Bragg peak effect of the proton irradiation and insure uniform whole-body exposure. Dose rates varied within groups but were less than 100 rad min^{-1} in all cases. The dose figures given indicates that calculated at the Bragg peak [8, 9].

3. Skin Effects

A frequently observed latent effect of exposure to less than 32 MeV protons is thickening or fibrosis of the skin. After the acute desquamation reactions subside, the latent fibrosis of the skin begins. In these cases thickening progressed to the point where the skin loses its elasticity. Attempt to handle these severely fibrotic primates result in tearing of the skin. Slowly healing lesions ensue which often become invaded by bacteria and fail to respond to treatment.

For primates exposed to 28 MeV protons, approximately three years was required after irradiation for extreme skin fibrosis to occur with 1000 rad exposure. Slight fibrosis or thickening was noted in animals two years after exposure with 1000 rad. With 5 MeV proton exposures, animals receiving 2000 rad show slight fibrosis two years after exposure and severe fibrosis by four years after. Exposure to 1500 rads will cause slight fibrosis within four years. A minimum of eighteen months is required for these latent skin effects to become evident. Primates which were not rotated during exposure and exposed directly to the chest and abdomen exhibited fibrosis somewhat earlier than rotated groups. Some minimal skin thickening has been noted in the remaining 32 MeV proton group, but only in the animals exposed to 560 and 280 rad does it remain after 7 years. For energies higher than 32 MeV the major effects have been seen in the deeper organs, though some areas of focal fibrosis with acanthosis have been found in skin biopsies. Primates exposed to 1800 and 1500 rad of solar flare simulating protons showed mild to moderate radiodermatitis within two years of exposure. Below 1200 rad the skin of these primates showed no gross changes of irradiation exposure. Earlier biopsies showed only minimal changes in the 1200 rad group and no changes in lower dose groups. The primates exposed to 12 MeV electrons show no skin changes below 900 rad on gross examination at approximately $1^1/_2$ years after exposure. Slight thickening of the skin was present in two of four primates at 1200 rad. Marked fibrosis was present in three of four primates in the 1500 rad group. Three years after exposure, the 1000 rad group of the primates exposed to 1.6 MeV electrons show no gross skin damage. The entire group of five primates exposed to 1500 rad show thickening of the skin with a loss of elasticity.

4. Testicular Effects

Latent effects are present in the testicles of the irradiated primates. Primates exposed to the lower energy protons have not reached sexual maturity and testicular changes cannot be assessed. With greater than 32 MeV protons and primates which have matured since exposure, testicular lesions have been noted, primarily involving decreased size and textural changes, i.e. small firm testicles. In the 55 MeV group, eight of ten animals with abnormalities appear upon examination and palpation to have one normal testicle and one damaged testicle. With energies greater than 55 MeV, 16 of 18 primates with abnormalities have involvement of both testicles. The method of exposure may be involved, however; for instance in the 55 MeV group shielding of one testicle by the primate's legs may have occurred. Fig. 1 shows a somewhat dose-related composite graph from the males with gross abnormalities within the 55, 138, 400 MeV and 2.3 GeV proton groups.

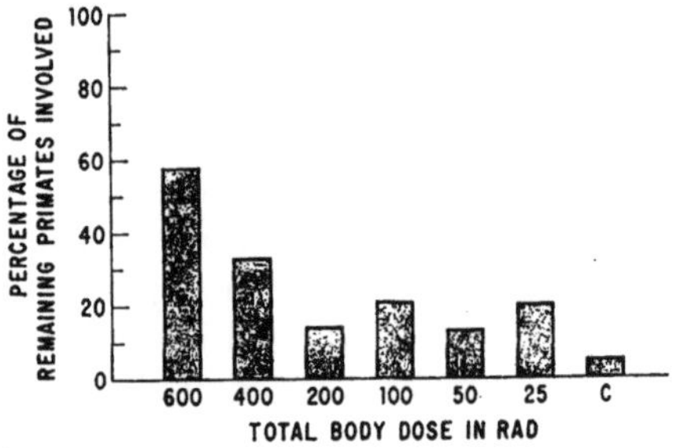

Fig. 1. Testicular abnormalities: composite for 55, 138, 400 and 2300 MeV protons.

Ten primates were selected as representative of the animals with small firm testicles and testicular biopsies were taken. These biopsies indicate that a fair percentage of atrophic, non-functional tubules occupy the smaller testicles. Actually, a graded effect was seen from atrophied to near normal tubules. Biopsy was also carried out on three non-exposed control animals for comparison with the irradiated primates. No atrophic tubules wee seen. Further studies are planned to elucidate the functional capacity of the animals with the atrophied testicles.

5. Cataractogenesis

Cataracts have been noted in large measure within the colony primates as the composite data for the 55 MeV–2.3 GeV proton groups illustrate (Fig. 2). As with the testicular composite curve, a dose effect relationship is present.

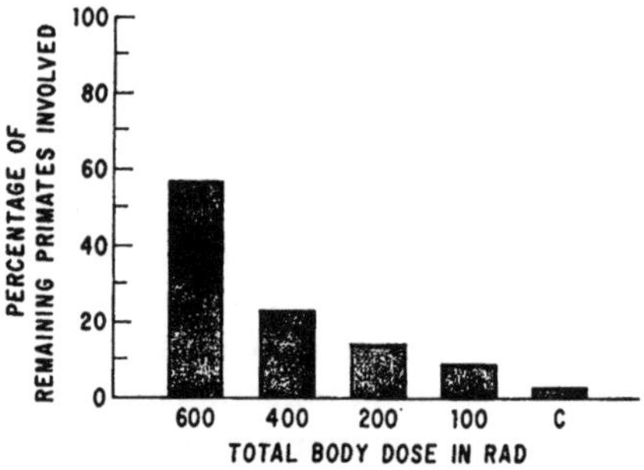

Fig. 2. Lens lesions in the eye: composite for 55, 138, 400 and 2300 MeV protons.

It should be noted, however, that these data represent lenticular lesions from the earliest tiny specks through focal opaque areas to the most severe cataracts. When the composite data are categorized into the several stages of possible cataract formation, the extent of true cataract formation can be ascertained (Figs. 3, 4). Fine bright specks which usually appear on the periphery of the lens are apparently not permanent lesions. At previous eye examinations

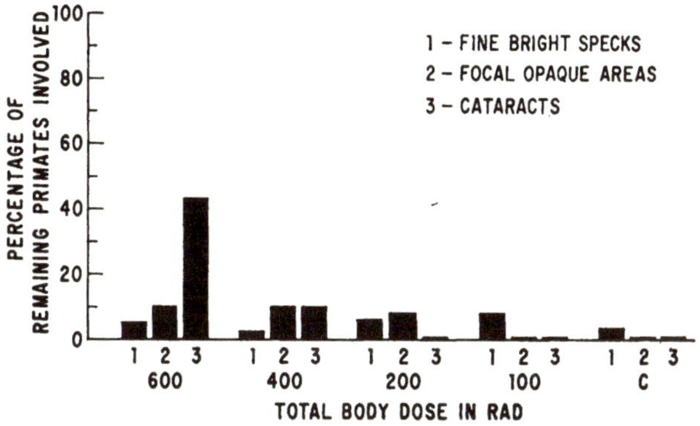

Fig. 3. Lens lesions in the eye: composite for 55, 138, 400 and 2300 MeV protons, analysed into fine bright specks, focal opaque areas and cataracts.

several specks were noted which upon more recent examination had disappeared. However, when these specks move to a more polar area in the lens, usually in the posterior subcapsular region, they come together to form focal opaque areas which are permanent but may either remain static or progressively increase in size until they cover the posterior subcapsular surface of the lens to form cataracts. With severe reactions, anterior subcapsular lesions may also form in a similar fashion to the posterior lesions but somewhat delayed.

Approximately six years after exposure, the lowest dose of penetrating protons (55 MeV or greater) necessary to produce a true cataract is 400 rad. Doses of 200 rad are clearly capable of producing focal opacities within the lens, and some specks which may potentially represent opacities have been noted with doses as low as 100 rad. Further eye examinations will reveal more about the cataractogenesis of these early lesions. The lower energy protons (5–28 MeV) are also capable of causing cataracts. The early lens lesions, such

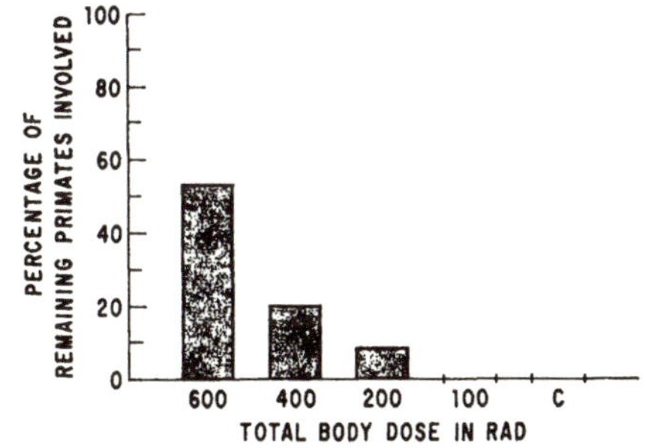

Fig. 4. Focal opacities and cataracts: composite for 55, 138, 400 and 2300 MeV protons.

as bright specks and focal opaque areas, have not been observed in primates as numerously as anticipated. In the 28 MeV group, cataracts are present in 12 of 13 animals, in half of them only three years after exposure at 1000 rad. However, some early lesions are present in the 10 and 15 MeV groups most noticeably two years after exposure at 2000 rad. The lack of early lesions suggests that either all the non-permanent lesions have been resolved, or when present, the early lesions rapidly progress to cataracts. Another point of interest is the 5 MeV group where two of two non-rotated primates have cataracts at the 3000 rad level. Three years after exposure, only one of 19 rotated primates, including the 3000 rad group, have lens lesions. The depth of penetration is less than 0.5 mm in tissue with 5 MeV protons, therefore limiting their entrance to direct eye exposure. By physical measurement, these protons should not reach the lens.

Of seven animals exposed to 1200 rad of solar flare irradiation, six have focal opacities in at least one eye, and three have bilateral opacities; only one animal had no lesion in either eye. Five of seven animals in the 900 rad group have focal opacities in at least one eye; four opacities were bilateral in these animals. In the 600 rad group, three of six animals have small focal opacities while three had no lesions. No definite opacities were seen in either the 300 rad or the control animals which number 14. $2^1/_2$ years after exposure, no eye lesions are present in a group of five primates exposed to 1000 rad of 1.6 MeV electrons. Two of five primates exposed to 1500 rad exhibit no lesions, whereas the other three have lens lesions in one eye only although whole-body exposure was made. Bright specks are present in two of the three and a small posterior subcapsular focal opaque area was noted in the third primate. In the 2 MeV electron exposed group, there are no marked lenticular lesions $1^1/_2$ years after exposure with whole-body doses of 1500, 1200 and 900 rad. No dose-related arrangement of the few bright specks can be made as nine of the sixteen animals have no lesions at this time. One animal has an early opaque area in a single eye but the remaining animals exhibit only small bright specks.

Of 29 primates remaining seven years after exposure to 2 MeV X-rays, only five animals have lenticular lesions. One of five animals receiving 716 rad whole-body exposure has early bilateral posterior subcapsular cataracts. Three of eight animals at 624 rad have early focal to diffuse opaque changes. One of ten primates in the 538 rad group has a small bright speck. No lesions are present in the 400 or 360 rad groups. The single remaining control primate has no lens lesions.

6. Hematology

Routine hematological and serum biochemical testing has proved to be a valuable screening procedure. The entire colony is examined quarterly. Values for the following are routinely ascertained:
Blood: total leukocyte count, differential count, microhematocrit, hemoglobin, platelet count, total red blood cell count.

Serum biochemistry: SGOT, alkaline phosphatase, glucose, BUN, total protein, albumin, total globulins, protein electrophoresis ($\alpha 1$, $\alpha 2$, β and γ fractions).

As a result of the hematological testings, a leukemic monkey which received 100 rad of 400 MeV protons was found 32 months after exposure. The monkey had total white blood cell counts in excess of 200000 cells per mm^3 and granulocytic leukemia was diagnosed. No other cases of leukemia have been found. Presently there are two confirmed cases of diabetes mellitus within the colony. Persistent hyperglycemia, glycosuria and abnormal glucose tolerance tests were manifested. One of the confirmed diabetics is in the 2.3 GeV group at 560 rad; the other animal received 446 rad of 2 MeV X-rays. There are at least ten other primates within the colony that also have elevated glucose levels. Urinalysis and glucose tolerance testing will be done on these animals in the near future. Additionally, there are numerous primates which have elevated SGOT levels with decreasing albumin levels and abnormal globulin fractions. Where indicated, needle liver biopsies are performed. Biopsy results include mild hepatic lipidosis, mild focal subacute hepatitis and abnormal nuclei similar to those seen with diabetes. The latter finding was in the diabetic primate exposed to 2.3 GeV protons.

7. Tumors

Several tumors have been noted within the irradiates but none have been found in the control population. Of particular interest are three brain tumors which have been characterized by Dr. Webb Haymaker, NASA Ames Research Center, as two malignant ependymal tumors and a malignant glioma [10]. These primates were in the 55 MeV group, two at 800 rad and one at 600 rad. The neoplasia became clinically apparent 3 years 8 months, 4 years 6 months, and 5 years after exposure. Two fibrosarcomas, one in the maxilla and the other in the thigh were found in another monkey. This monkey had received 600 rad of 55 MeV protons and was sacrificed 4 years and 5 months after exposure. The thigh tumor developed to a size of 10–12 cm in diameter in a period of about 60 days after becoming clinically apparent. Approximately two years postexposure, a fibroma of the abdominal skin was noted in a solar flare exposed primate. This animal received 1500 rad. The tumor was about $1^1/_2$ cm in diameter and appeared to be benign. An additional fibrosarcoma was found just below the knee joint in a 1.6 MeV electron exposed primate. The tumor appeared approximately three years after exposure to 1500 rad. After suddenly appearing, it grew to 3 inches by 2 inches in size within two weeks after appearance. Some bone involvement was visualized on diagnostic radiography.

8. Deaths

Deaths from a variety of conditions in addition to the tumors have occurred in the past two years. These included Shigellosis, chronic interstitial nephritis, five cases of acute gastric dilatation, peracute hemorrhagic pancreatitis, generalized amyloidosis, chronic cystitis, chronic active diverticulitis, streptococcal septicemia and endometriosis. Radiation related degenerative changes noted on post-mortem examination include; thyroid atrophy; mild to severe chronic radiodermatitis; muscular atrophy, especially of the limbs; fibrosis of skin and muscle across joints restricting movement; superficial scarring on the heart, lungs, liver, adrenal glands and other organs; and stenosis of the intestines due to scarring. A common finding in all the primates is the presence of lung lesions due to the monkey lung mite.

9. Body Weights

In a survey study of the 2 MeV X-ray, 55–400 MeV, and 2.3 GeV proton groups, a difference was found in the body weights between the exposed and control animals only in the 55 MeV group. Groups exposed to all other energies (MeV) showed no body weight difference from the controls or between groups. All the 55 MeV proton irradiated groups had smaller body weights than the control group, where six primates averaged 10.50 kg. The greatest difference was for the 600 rad group which averaged 6.77 kg for eight remaining primates. The other groups were scattered between these two extremes but not in a dose-related fashion. Body weights, available back to the exposure event, are being statistically analyzed to determine when significant differences in body weight became evident.

10. Alpha Experiences

The final group to be mentioned is a small group of primates exposed to alpha particles. One group received single eye exposures of 108 and 85 MeV alpha particles in dose ranges from 250 to 1000 rad at 100 rad min^{-1}. No lesions have been noted using biomicroscopic slit lamp techniques in the 22 animals, each serving as its own control. An additional twelve primates were exposed to 108 MeV alpha particles in 1000, 1500 and 2000 rad total doses to an area of skin 49 cm^2 located on the abdomen below the sternum. To date no gross desquamation changes as seen in the skin penetrating proton exposures have been noted. Skin biopsies taken 32, 100 and 180 days after exposure show moderate acanthosis, mild focal epithelial dysplasia with mild dermal fibrosis and vasculitis of subepithelial vessels. None of these changes can be related to the magnitude of the radiation dose. Four animals trained to a visual acuity test and having both eyes exposed to 1000 rad of 108 MeV alpha particles have shown no learning decrement.

11. Conclusion

In summary, the most clear-cut radiation effect which appears to be detectable down to 200 rad after 6 years and is dose related is cataractogenesis. Carcinogenic effects have not been marked to date, but all neoplasms to date have occurred in radiated primates and radiation-induced neoplasms in primates may have long latent periods. Proton energies that penetrate only the superficial portions of the body produce severe fibrosis with doses of 800–1000 rad and upwards in approximately three years. Skin changes with doses of 600 rad or less have been insignificant. Recently a number of the screening tests have shown alterations in some biochemical parameters. Whether these are related to radiation effects is not known at this time but may be an expression of the non-specific life shortening effects frequently attributed to radiation.

Certificate

The animals involved in this study were maintained in accordance with the "Guide for Laboratory Animal Facilities and Care" as published by the National Academy of Sciences, National Research Council.

(*signed*) John H. Kirk, *Capt. USAF, VC*

Acknowledgments

The research reported in this paper was conducted by personnel of the Radiobiology Division, USAF School of Aerospace Medicine, Aerospace Medical Division, AFSC, United States Air Force, Brooks AFB, Texas. Further reproduction is authorized to satisfy the needs of the US Government.

References

[1] G. V. Dalrymple and I. R. Lindsay, Radiation Res. 28, 365 (1966).
[2] W. H. Langham, Radiobiological Factors in Manned Space Flight, National Academy of Sciences, Washington, D. C., 1967.
[3] G. V. Darlymple et al., Radiation Res. 28, 406 (1966).
[4] G. V. Dalrymple et al., Radiation Res. 28, 434 (1966).
[5] I. R. Lindsay et al., Radiation Res. 28, 447 (1966).
[6] G. V. Dalrymple et al., Radiation Res. 28, 471 (1966).
[7] G. V. Dalrymple et al., Radiation Res. 28, 507 (1966).
[8] G. H. Williams, J. D. Hall and I. L. Morgan, Radiation Res. 28, 372 (1966).
[9] J. C. Mitchell et al., Radiation Res. 28, 390 (1966).
[10] W. Haymaker, personal communication, March 1971.

Life Sciences and Space Research X — Akademie-Verlag, Berlin 1972

ANALYSIS OF SURVIVAL AND CAUSE OF DEATH STATISTICS FOR MICE UNDER SINGLE AND DURATION-OF-LIFE GAMMA IRRADIATION

D. Grahn, R. J. M. Fry and R. A. Lea

Division of Biological and Medical Research,
Argonne National Laboratory, Argonne, Ill., USA

The late effects of protracted exposure to low levels of external radiation continue to be a matter of operational concern in long-range space flight. Studies have been carried out on young adult mice exposed to daily levels of ^{60}Co gamma irradiation ranging from 0.3 to over 30 R day^{-1}. The lowest level is comparable with the occupational maximum permissible dose for the atomic energy industry. There is little evidence of life shortening at that level, but as exposure increases, there is an exponential decline in life expectancy. The life-shortening coefficient is approximately 4 days/100 R accumulated or 4% R^{-1} day^{-1}.

When life-shortening is $\leqq 15\%$, all of the increased mortality can be attributed to radiation-induced increases in death rates from neoplastic diseases, including various forms of leukemia and pulmonary tumors. Age-specific death rates for mice dying of all other causes remain the same as the controls throughout life, at the lowest doses. A non-neoplastic disease component of excess mortality rate emerges at 6 R day^{-1} and above. The risk of death from all and specific causes following single exposures compared with protracted lifetime irradiation shows a clear effect of protraction. Leukemia death rates are reduced by a factor of 5 or more at all daily exposure levels below 20–30 R day^{-1}. Risks for other causes of death are also reduced, but to a variable degree.

1. Introduction

The chronic effects of whole-body exposure to penetrating ionizing radiations have been reasonably well characterized for mice in terms of life shortening and leukemia incidence following single doses of 100 R or greater and lifetime exposures averaging 5 R day^{-1} and above. Recent reviews and reports have summarized the state of our knowledge [1–3], and have also discussed the validity of life-shortening extrapolations from the laboratory mouse to man. While several extrapolation procedures are available, the ultimate basis for any such procedure relies on the basic similarities or analogies of the actuarial or vital statistics of the species compared. The expression of radiation injury in terms of changes in vital statistical parameters has been amply demonstrated [3–5] and, for life shortening, has been presented in quantitative dose-response relationships. Whether related to the radiation exposure of flight crews during long-range space missions or to the general population subject to a potentially rising exposure to environmental and industrial sources of radiation, the problems now at issue are those involving the prediction of specific pathological effects, of age-specificities, and the significance of variable dose-rate and exposure patterns. In this regard, the effects of very low levels of exposure, those within 100 times the average natural background level for the general population, still remain somewhat unresolved.

This paper is concerned with several aspects of these problems, as they can be approached through the use of vital statistical procedures. The general statistical methodology has been described in other reports [3–5]. Basically, the age-specific death rates are derived for the lesions of concern. The data are then fitted by least squares methods to the polynomial equation which minimizes the residual variance and which is also compatible with other data in the given comparison. Equations describing the age changes in death rate for an intercomparison group are found by fitting the data through a selected common death-rate value at the initiation of the experimental period from which all the dose-group populations originated. The curves of populations at different radiation levels can be differentiated with respect to that of the control population to provide a measure of the integrated lifetime excess risk for any given endpoint. This is generally equivalent to the actuary's mortality ratio.

The data are all derived from inbred and hybrid groups of mice of both sexes subjected to whole-body external ^{60}Co gamma irradiation. Single exposures were given at 100 ± 20 days of age, and lifetime daily irradiations were begun at the same age. Details of the exposure facility and conditions are given elsewhere [1, 3]. The major endpoints of these experiments have been survival time or mean after-survival (MAS) from the start of exposure, the major neoplastic diseases, including reticular tissue tumors (leukemia), pulmonary

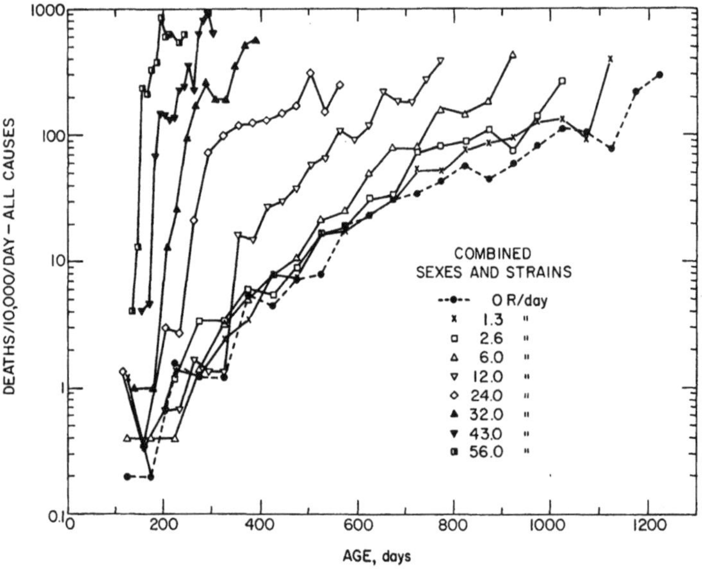

Fig. 1. Age-specific mortality rates for all causes of death for mice under daily ^{60}Co γ radiation started at 100 days of age. Data derived from the pooling of inbred strains A, BALB/c, C57BL/6 and the BCF_1 hybrid.

tumors, ovarian tumors and cysts, collectively all other tumors, and non-neoplastic diseases, which include renal diseases, pneumonia and other infectious conditions, and other known and unknown causes.

2. Results of the Analysis

The MAS for mice exposed for the duration-of-life to high energy γ-radiation declines exponentially with increasing levels of daily exposure. The data fit the general equation $MAS_D = MAS_0 \exp(-\beta D)$, where D is the daily exposure

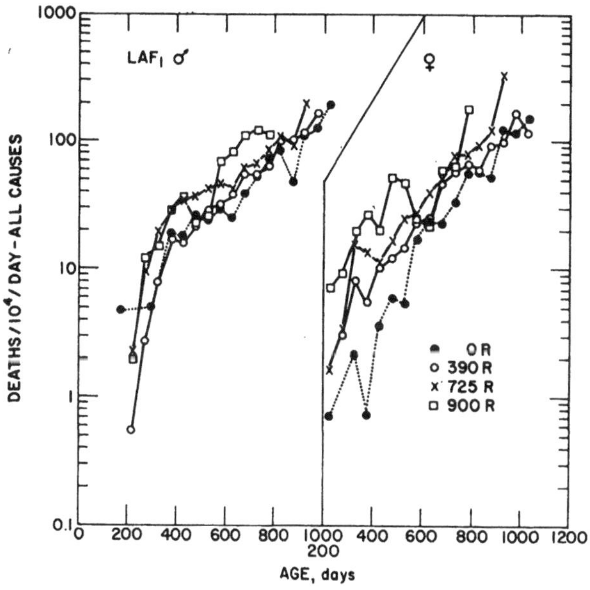

Fig. 2. Age-specific mortality rates for all causes of death for LAF_1 hybrid mice following single exposure to ^{60}Co γ radiation at 100 days of age.

level and β is the regression of ln MAS on D; β has the average value of -0.038 or about 4 days of life shortening per 100 R accumulated at low daily exposures. Exposure to daily irradiation causes the regression of age-specific death rate on age to diverge progressively from the control, as in Fig. 1, for all causes of death. The typical pattern for single doses is less clear. Ideally, and theoretically [4, 5], the displacement of the death rate regressions should be linear, that is, the death rates would be higher than the control and run parallel to that group. In fact, the regression may be non-linear with some convergence with the control late in life. The early post-irradiation period may be characterized by a non-linear phase of excessive mortality from acute leukemia. Fig. 2 demonstrates some of these characteristics for hybrid mice.

2.1. Age-specific Death Rate Analysis for Neoplastic and Non-neoplastic Diseases

The subdivision of the cause-of-death data into the neoplastic and the non-neoplastic disease death rates is of fundamental concern. The term "non-specific life shortening", which can logically be related to excess mortality from non-neoplastic disease, has found common usage, though it has always lacked a clear relation to pathological entities. Neoplastic conditions, on the other hand, have always attracted the analytical attention of those concerned with the establishment of radiation safety standards [6–8], and existing data on irradiated human populations have emphasized cancer as an endpoint of major significance [6, 8]. Fig. 3 presents a series of data on animals irradiated

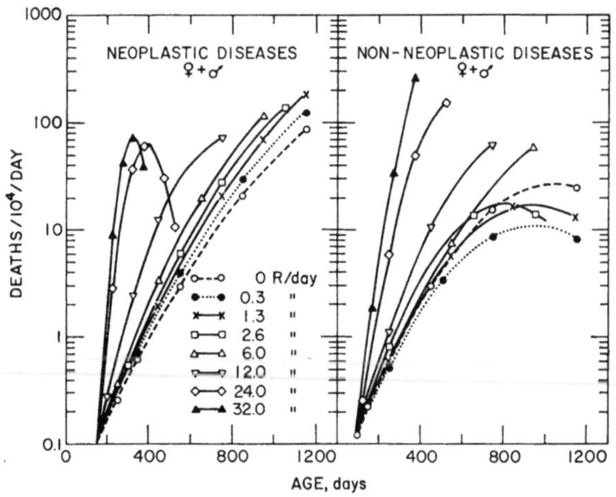

Fig. 3. Age-specific mortality rate, for indicated causes, fitted through common intercepts. Data from mouse strains BALB/c, C57BL/6 and their F_1 hybrid.

for the lifetime. The fitted curves reveal the existence of a steadily increasing increment of mortality associated with neoplastic disease as both age and daily dose-rate increase. At levels of 24 R day^{-1} and 32 R day^{-1}, almost all the tumor mortality is due to acute leukemia, but below these dose levels, a broad spectrum of tumor types is normally observed. The death rate data for non-neoplastic diseases shows a striking similarity among the regressions at all of the lower doses, with even some degree of depressed non-tumor death rate below the controls. There is no excess mortality from non-neoplastic conditions relative to the control at 2.6 R day^{-1} and below. At 6 R day^{-1}, a significant deviation from the control and the lower doses begins at about 600 days of age (500 days of daily exposure). This excess non-tumor mortality steadily increases with both age and daily level of irradiation.

Although these data lack the refinement of definition of "cause of death", the gross separation of the mice into the two classes, those dying with and those dying without evidence of tumors, is nevertheless a reliable differentiation. It is important to emphasize that the risks of early or excess death from radiation exposure at extremely low intensities, those within 100–200 times background, all appear to be related to the risk associated with neoplastic disease. The expression of these results in terms of MAS is complicated for two reasons; there is an overall increase in the risk of death from tumors under irradiation, and this is accompanied by a shift in the time of appearance of these deaths. Thus, as seen in Fig. 4, tumor deaths increase in incidence and the modal frequency shifts to an earlier age. The ln MAS for the all-tumors class declines linearly with increasing exposure level at a rate of about 4% R^{-1} day^{-1}.

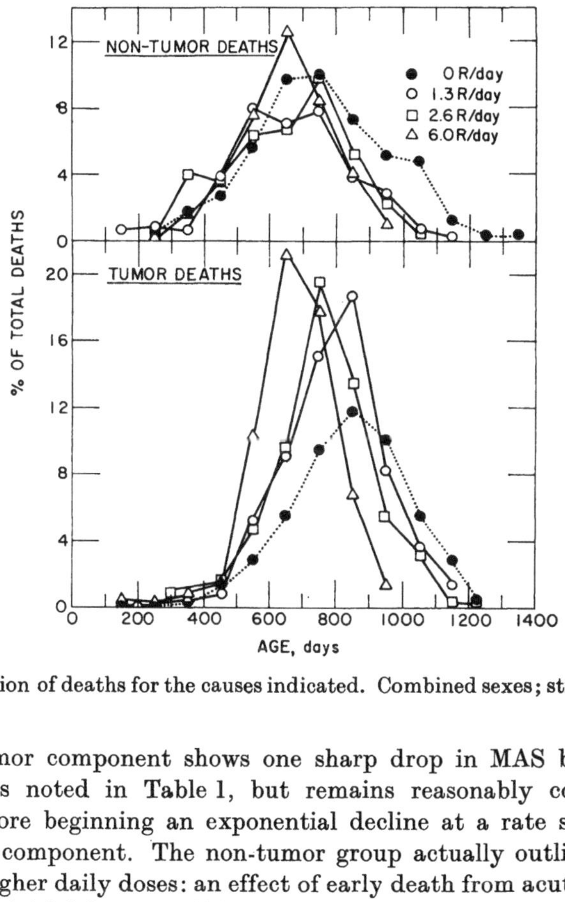

Fig. 4. Distribution of deaths for the causes indicated. Combined sexes; strains as in Fig. 3.

The non-tumor component shows one sharp drop in MAS between 0 and 1.3 R day^{-1}, as noted in Table 1, but remains reasonably constant up to 6 R day^{-1}, before beginning an exponential decline at a rate similar to that for the tumor component. The non-tumor group actually outlives the tumor group at the higher daily doses: an effect of early death from acute lymphocytic leukemia. The initial drop in MAS associated with the non-neoplastic diseases at 1.3 R day^{-1} can be explained by the nature of the non-neoplastic disease class. Although there is no increase in age-specific risk or probability of death from the non-neoplastic causes, the proportion of deaths from these causes is reduced from about 50% to 37% of the population, for the combined sexes, and most of this change is late in life. To remain in the non-tumor class, the animal must die before the appearance of a tumor, as the latter event irreversibly transfers the animal to the tumor class. Because tumor incidence increases

Table 1

Mean after-survival data and final total tumor incidences for the two sexes combined. Data from BALB/c and C57BL/6 mouse strains and their F_1 hybrid.

Daily level of exposure (R day^{-1})	Mean after-survival ± SE (days) Tumor deaths	Non-tumor deaths	Total deaths	Tumor incidence (%)
0	739 ± 10	653 ± 12	696 ± 8	51
1.3	688 ± 8	558 ± 13	640 ± 7	64
2.6	657 ± 11	546 ± 15	612 ± 9	60
6	577 ± 8	548 ± 12	565 ± 6	61
12	444 ± 8	444 ± 8	443 ± 5	53
24	245 ± 5	290 ± 7	268 ± 4	48
32	186 ± 3	202 ± 6	194 ± 3	46

with both time and total dose, the probability of remaining in the non-tumor class decreases steadily with age. The statistical effect is to shift the MAS downward by about 100 days. Although this could be interpreted as evidence for a non-specific component of life shortening, the fact that the MAS remains somewhat unchanged for the three lowest exposure groups raises doubts about that interpretation. The preferred interpretation is to consider this initial drop in MAS for non-tumor deaths to be something of a statistical artefact due to the rise in risk of death from tumors. The absence of an excess age-specific death rate for non-tumor causes at the lowest three doses is the significant statistical finding.

Findings similar to those described for lifetime irradiation are also found for mice subject to single doses at 100 days of age. Fig. 5 shows the compar-

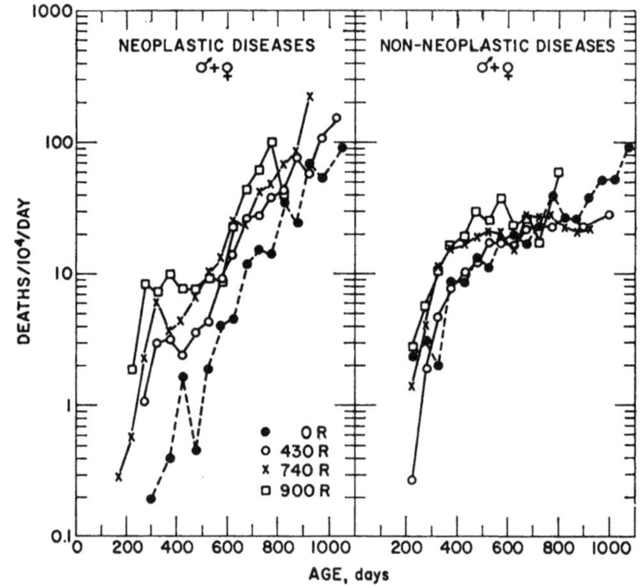

Fig. 5. Age-specific mortality rates for the causes indicated; LAF_1 hybrid mice.

ative data on age-specific death rates for the two subgroups: neoplastic and non-neoplastic diseases. In general, the non-tumor component has a non-linear regression on age, with the death rates becoming very steady or unchanged beyond $1^1/_2$ to 2 years of age. The initial rapid rise in non-tumor death rate is concurrent with that seen for the tumor group. These may be interrelated in that the pre-leukemic and early leukemic states are often characterized by increased infectious disease susceptibility. Unexpectedly, the non-tumor death rate continues to rise beyond the tumor death rate, when the latter drops back after the initial phase of acute lymphatic leukemia subsides. After this initial phase, the rate of mortality from all tumors rises steadily throughout life and remains elevated above the control. A small dose-differential prevails.

The MAS data are also similar to those seen under daily irradiation in that the principal life-shortening effect is attributable to excess tumor mortality, which increases from 32% to 46% of all deaths between 0 R and the lowest single dose group (Table 2). The MAS for these low dose non-tumor deaths drops 50 days from the controls, but again, as for the low dose daily irradiation series, this is a statistical effect of changes in the distribution of the two major causes of death, and it is not associated with an excess risk in terms of age-specific mortality. There is, however, some excess risk for the two highest single dose groups in this non-tumor category of death.

Table 2

Mean after-survival data and final tumor incidences for LAF_1 hybrid mice following single exposures to ^{60}Co γ-rays. Sexes combined.

Dose	Mean after-survival ± SE (days)			All
	Tumor	Non-tumor	All	Tumors
(R)	deaths	deaths	deaths	(%)
0	692 ± 12	532 ± 11	582 ± 8	32
390	614 ± 10	480 ± 9	542 ± 7	46
725	544 ± 11	403 ± 9	464 ± 7	44
900	462 ± 20	384 ± 15	415 ± 12	40

2.2. Analysis of the Effects of Dose-Protraction on Excess Mortality

Although radiologists and therapists have long recognized the importance of dose fractionation and protraction to reduce the damage potential to normal tissues of a given exposure series, there currently seems to be some doubt that a "protraction factor" truly exists [9]. By "protraction factor" we mean the reduced biological effectiveness that is observed for a given endpoint when a radiation exposure is delivered at a low intensity over a long period, compared with an exposure given at high intensity over a short time period. The death rate analysis, and the integral and differential excess mortality, relative to the control, provide a quantitative approach to this problem. Fig. 6 gives the age-specific leukemia death rates for a hybrid mouse, the LAF_1 female, exposed to ^{60}Co γ-radiation. The single doses ranged from 150 R to 1100 R, but were pooled into three subgroups to increase the sampling statistics.

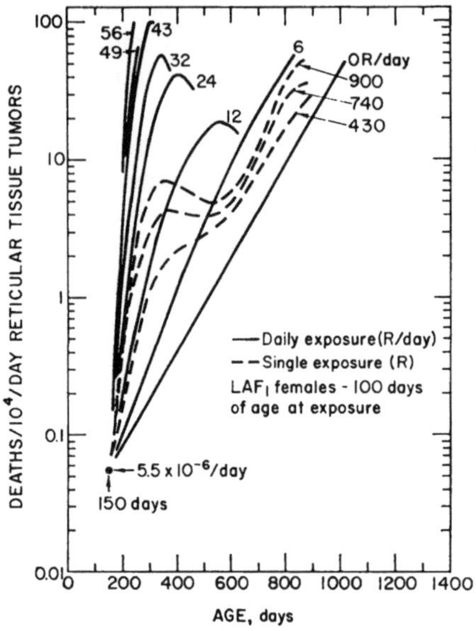

Fig. 6. Mortality rate data for all leukemias, LAF_1 females, fitted through common intercept.

The typical early phase of acute leukemia occurs at all single-dose levels and also at the higher daily exposures. The final cumulative incidences of leukemia are given in Table 3. A comparison of the incidence data in Table 3 and the death rate data of Fig. 6 clearly reveals the non-informative nature of simple incidence data in contrast to the descriptive sensitivity of the probabilistic analysis to age, dose and dose rate factors in radiation leukemogenesis. A series of integral mortality ratios is plotted against accumulated dose in Fig. 7 for the daily dose data. A single lifetime ratio is derived for the single dose series. It is clear from these data that all the lower daily doses, 32 R day^{-1} and below,

Table 3

Final leukemia incidences for the LAF_1 female under daily or single exposures to ^{60}Co γ radiation

Dose	No. of mice	% leukemia
0 R	262	20.6
5 R day^{-1}	120	25.8
12 R day^{-1}	183	25.7
24 R day^{-1}	181	31.0
32 R day^{-1}	150	31.3
43 R day^{-1}	145	23.4
49 R day^{-1}	135	12.6
56 R day^{-1}	118	5.9
390 R	369	23.8
725 R	348	25.6
900 R	110	25.5

conform to one trend of rising leukemia risk with increasing accumulated dose. Though daily dose rates vary from 5 R to 32 R, the instantaneous dose rate in all cases is less than 30 milliroentgens per minute. In contrast, the single doses, delivered at 2–20 R per minute, give an excess risk for leukemia mortality that is at least five times greater than for comparable levels of accumulated dose under low-intensity daily irradiation.

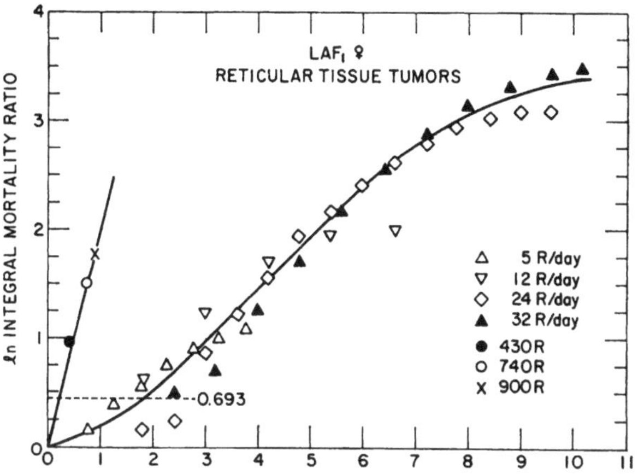

Fig. 7. Integral mortality ratios as related to dose, for LAF_1 females, derived from the unirradiated control (see Fig. 6).

Admittedly, there are confounded time, dose and death rate variables. At 5 R and 12 R per day total doses of 400 R–900 R are accumulated in a time period of only 30–180 days. The time of maximum differential death rate (see Fig. 8) occurs at 200 days after exposure for all the single doses, while the times

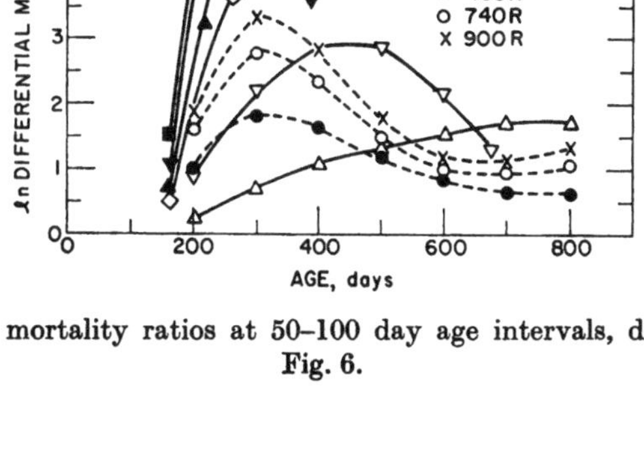

Fig. 8. Differential mortality ratios at 50–100 day age intervals, derived from data in Fig. 6.

of maximum differential under 12 R and 5 R per day occur at 450 and 750 days respectively. These latter ages are significantly more than 200 days beyond the age point where total dose accumulations are equivalent to the compared single exposures. Thus, the latent period, or early tumor development stage, must be greatly increased under low intensity irradiation. In addition, there is a general decrease in the risk of leukemia induction per R of exposure at 5 and 12 R day^{-1}. Although the final cumulative leukemia incidences are equal to those at 725 R and 900 R, the final cumulative doses exceed several thousand roentgens even when discounting the last several hundred days of exposure to permit equal time for the leukemia to be manifest. The protraction factor, therefore, seems a genuine effect of radiation intensity although one variable does remain to be isolated. The daily exposures did occur over an age interval of 1–5 months, during which period the single exposure groups were allowed full expression of any induced injury. To eliminate the possibility that this brief time period of five months was one wherein a sharp decline in sensitivity to radiogenic leukemia might have occurred, a separate experiment would be required. In such an experiment, the single exposures would have to be given at several age levels between 100 days and 250 days of age, and the daily exposures of 5 and 12 R day^{-1} would have to be terminated at appropriate times and total doses within the same time interval. Until this is done, the claim can still be made that all the protraction effect is attributable to the age differences over which the doses were delivered. While it seems obvious that such would not be the case, the criticism must eventually be answered.

The protraction factor must also be more thoroughly investigated for both high and low LET (linear energy transfer) radiations, as it is a critical factor for long range mission planning in space operations. The galactic radiation component, for example, while small in terms of dose per day, does contain a high LET component whose full effect is not predictable. Overall, the galactic radiations contribute a steady increment of radiation exposure. Other exposure factors, as solar proton events, would generally be infrequent and of both low intensity and low total dose behind most anticipated shielding situations for flight modules. Thus, mixtures of fractionated and protracted, high and low LET, radiations will characterize the exposure parameters of all long range flight missions. Improved knowledge of the degree of additivity of separate exposures to high LET radiations and the extent to which protraction and fractionation will reduce the effectiveness of low LET radiations is imperative for better planning. Data on all causes of death other than leukemia, seen in Fig. 9, show that the protraction factor for non-leukemic deaths has a somewhat different relationship to dose-rate. As the daily rate rises toward 40–50 R day^{-1}, though still at intensities of less than 50 milliroentgens per minute, the sparing effect of low intensity exposure drops from a value of 5 at 12 R day^{-1} to less than 2 at 40–50 R day^{-1}. This may reflect a steady increase in hematopoietic system injury, for example, that may be expressed as a disproportionate increase in death probability from infectious disease.

How such data will translate to man is another problem that must yet be rationalized. For the moment, daily exposure rates for man in space flight

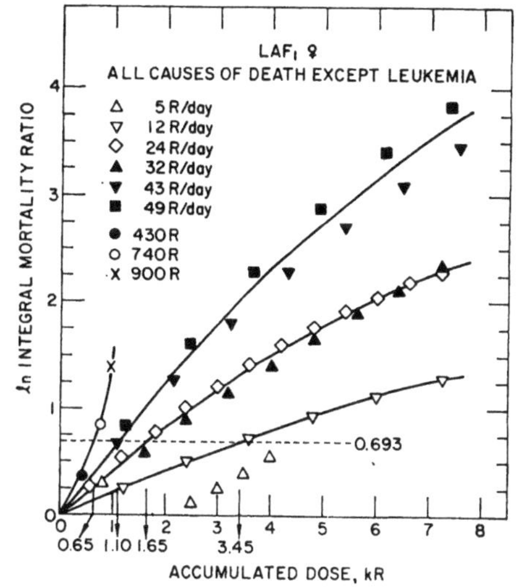

Fig. 9. Integral mortality ratios derived from unirradiated control for all causes of death except leukemia, LAF_1 females.

should be conservatively held below 1 R day^{-1}, preferably below 0.5 R day^{-1} to the bone marrow. These are the recently published recommendations of the Radiobiological Advisory Panel, Space Science Board, US National Academy of Sciences, National Research Council [10]. The present data fully support the general recommendations of that Advisory Panel, and, in addition, substantiate the existence of a significant protraction factor. This factor has a value of about 5 for leukemia in the mouse at levels up to 30 R day^{-1}, but for all other causes of death, the factor of 5 appears limited to doses below 10 or 15 R day^{-1}. At higher intensities, the sparing effect is reduced. With further experimentation, it should be possible to isolate the cause-specific protraction factors in greater detail and identify special problems that may relate to the high LET galactic radiations.

Acknowledgment

This work was supported by the US Atomic Energy Commission.

References

[1] G. A. Sacher and D. Grahn, J. Natl. Cancer Inst. **32**, 277 (1964).
[2] D. Grahn and G. A. Sacher, in: USAEC Report CONF-680410, Washington 1968 (pp. 2.1–2.27).
[3] D. Grahn, in: Late Effects of Radiation, Taylor and Francis Ltd., London 1970 (p. 101).
[4] G. A. Sacher, in: Radiation and Ageing, Taylor and Francis Ltd., London 1966 (p. 411).

[5] G. A. SACHER, Radiology **67**, 250 (1956).
[6] International Commission on Radiological Protection, Radiosensitivity and Spatial Distribution of Dose, Publication No. 14, Pergamon Press, New York 1969.
[7] Radiobiological Factors in Manned Space Flight, US National Academy of Sciences, National Research Council, Washington, D. C. 1967.
[8] United Nations Scientific Committee on the Effects of Atomic Radiation, General Assembly, Supplement No. 14 (A/5814) United Nations, New York 1964.
[9] A. R. TAMPLIN and J. W. GOFMAN, Population Control through Nuclear Pollution, Nelson-Hall, Chicago 1970.
[10] Radiation Protection Guides and Constraints for Space-Mission and Vehicle-Design Studies involving Nuclear Systems, US National Academy of Sciences, National Research Council, Washington, D. C. 1970.

Life Sciences and Space Research X — Akademie-Verlag, Berlin 1972

BIOLOGICAL EFFECT OF COSMIC AND TELLURIC RADIATIONS

H. PLANEL, J. P. SOLEILHAVOUP and R. TIXADOR

Laboratory of Medical Biology, University of Toulouse, Toulouse, France

Paramecium caudatum and *Paramecium aurelia* cultures were placed in lead shielding devices 5 and 10 cm thick. A diminution of growth rate was observed even for cultures carried out in sealed ampoules. Normal growth is restored when shielded cultures are irradiated by very low doses of gamma radiations (dose rate 700 mr/year). Irradiations by ^{60}Co (dose rate ranging from 600 to 39000 mr/year) without shielding are followed by an acceleration of growth rate. The same results have been observed in an underground laboratory. Measurement of natural radiation intensity was performed by gamma spectrography.

In summary, these results have successfully demonstrated the biological effect of background radiations and shown that these radiations may activate cell multiplication.

To be published in full in the *International Journal of Radiation Biology.*

Effects of Space on Living Matter

Life Sciences and Space Research X — Akademie-Verlag, Berlin 1972

EFFECTS OF SIMULATED SPACE VACUUM ON BACTERIAL CELLS

H. BÜCKER, G. HORNECK, R. FACIUS, M. SCHWAGER, C. THOMAS, G. TURCU and H. WOLLENHAUPT

Arbeitsgruppe für biophysikalische Weltraumforschung, Universität Frankfurt, FRG

The effect of vacuum on bacterial cells is related to water desorption. Below water vapour pressure the inactivation remains constant, independent of total pressure and exposure time. In subsequent growth, the lag-phase of the survivors is delayed. Combined treatment with vacuum and radiation (X-rays or uv of 254 nm wavelength) results in synergistic effects, whereas vacuum and heat can act antagonistically. The vacuum inactivated cells (indicated as loss of colony-forming ability) are completely damaged. They do not show cellular elongation, phage production or respiration. The cellular membrane becomes permeable by vacuum exposure: biomolecules are released from the cells when re-suspended after vacuum treatment.

The biological effect of vacuum up to ultra high vacuum is being studied on bacteria. One aim of these investigations is to find the mechanism by which vacuum affects biological matter. *E. coli* and *Bac. subtilis* vegetative cells and spores are being used. It has been shown that the inactivation regarded as loss of colony-forming ability is related to water desorption: during evacuation the colony-forming ability decreased rapidly as soon as the total pressure was decreased below the vapour pressure of water [1]. The survival of unprotected cells was about 5% for *E. coli B/r* and vegetative cells of *Bac. subtilis*, and about 20% for *Bac. subtilis* spores. Beyond this step in the range of water vapour pressure the degree of inactivation did not depend on pressure, which was tested up to 10^{-9} torr. This result is supported by theoretical considerations [2]. Under the assumption that the process of inactivation may formally be considered as a single chemical reaction for which a reaction rate can be defined, it can be expected from thermodynamic properties (Boltzmann factor) that there should be practically no difference in survival from 10^{-6} torr up to 10^{-17} torr.

Although in regard to survival no revolutionary results may be expected in the pressure range of space vacuum, it is necessary to conduct experiments at lower pressure, in order to get more information about the mechanism, for instance by sensitive detection of desorbing molecules and by studying molecular surface reactions. Furthermore no dependence on exposure time to vacuum was detected for survival up to three months. However, in case of "protected" exposure the survival decreased with time (Fig. 1). This protection was achieved by exposing the bacteria in suspension of phosphate saline instead of distilled water. After an exposure time of 80 days the survival of such protected spores of *Bac. subtilis* approaches the values for unprotected exposure.

In addition to tests on survival, indicated as colony-forming ability, further experiments have been performed to answer the following questions:

1. How are the colony formers affected by vacuum?
2. To what degree are the non colony formers injured?
3. How do vacuum induced injuries become manifest?

1. *How are the colony formers affected by vacuum?*

This question has been discussed already in earlier papers [1, 3–6]. The results we obtained up to now are summarized as follows:

(i) The first cellular division of vacuum exposed cells is delayed compared with the controls. The lag-phase is prolonged for about 45 minutes. This delay

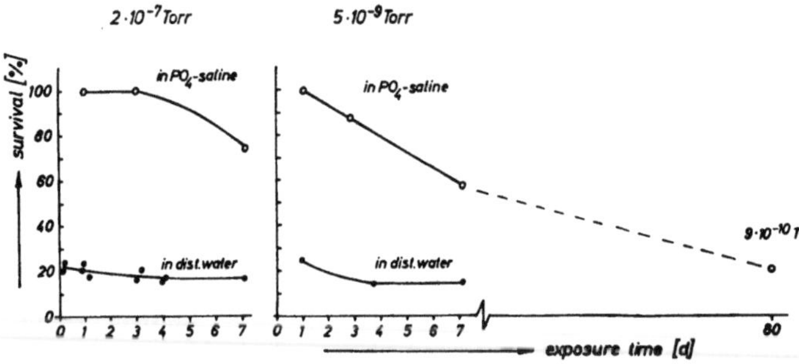

Fig. 1. Survival of *Bac. subtilis* spores exposed in distilled water or PO_4-saline to vacuum of 10^{-7} torr or 10^{-9} torr.

does not depend on final pressure in the range considered, about 10^{-1}–10^{-6} torr. Therefore the water desorption of the cells is assumed to be the reason for growth delay. Experiments will be conducted in order to test which processes are responsible for this effect.

(ii) The radiation sensitivity is increased in vacuum. Studies on *E. coli* showed the slope of the inactivation curve to be increased by a factor of 4 for X-irradiation and by a factor of 6 for uv irradiation. Nearly the same supersensitivity to uv in vacuum was observed on vegetative cells and spores of *Bac. subtilis*. The lethal uv lesions produced in vacuum are not photoreactive, in contrast to those of the controls. On the other hand the photoreactivating enzyme was shown to be intact after vacuum exposure. Furthermore the lethal uv lesions obtained in vacuum are not, or only a little, accessible to excision repair. From this it is inferred that the inactivation by uv in vacuum is not due to pyrimidine dimers, but possibly to spore type products or DNA protein crosslinks. Investigations on the nature of the photoproducts obtained in vacuum are being performed.

(iii) The sensitivity to heat (80 °C) was reduced in vacuum for spores of *Bac. subtilis*.

2. *Two what degree are the non colony formers injured?*

The loss of ability to form macrocolonies had been taken as an indicator of inactivation, i.e. loss of ability for growing out to about 10^8 cells. Therefore the question arose: what occurs to the individual cells just after vacuum exposure? Do those cells that we consider to be inactivated because they are unable to grow out to macrocolonies still show any metabolic or reproductive functions?

After vacuum exposure the growth of *E. coli B/r* in nutrient broth was observed. The colony formers on nutrient broth agar plates were counted periodically. In parallel the following tests were conducted: (i) Cellular elongation, the first visible indication of growth, was observed microscopically. It

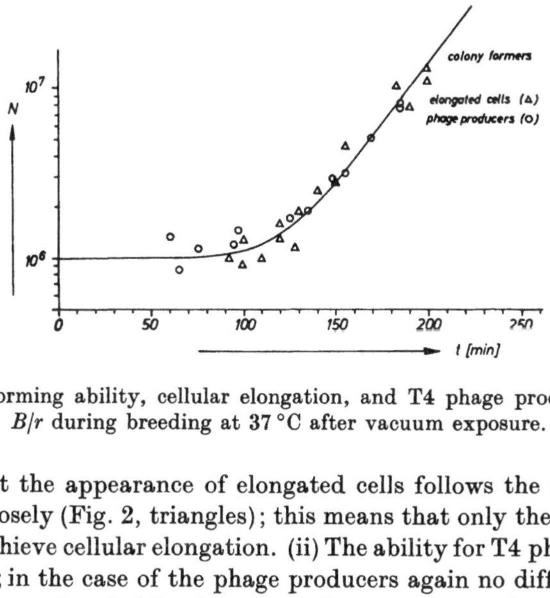

Fig. 2. Colony forming ability, cellular elongation, and T4 phage production of *E. coli B/r* during breeding at 37 °C after vacuum exposure.

was shown that the appearance of elongated cells follows the curve of colony formers very closely (Fig. 2, triangles); this means that only the colony formers were able to achieve cellular elongation. (ii) The ability for T4 phage production was also tested; in the case of the phage producers again no difference with the colony formers was found (Fig. 2, circles). In comparison, Cox and Baldwin [7] found in *E. coli*, air dried at 25% relative humidity, twice as many elongated cells and T7-phage-producers as colony formers. (iii) The respiration of the survivors was tested by means of oxygen consumption in a Warburg apparatus. It has been shown that only the colony formers were capable of respiration.

From these results it follows that the vacuum inactivated cells (as indicated by loss of colony forming ability) are completely damaged. They are unable to resume basic metabolic or reproductive functions such as cellular elongation, phage production or respiration.

3. *How do vacuum induced injuries become manifest?*

It has been shown that inactivation occurs during water desorption of the cells. But what processes are exactly responsible for this damage? Has the cellular membrane been affected by vacuum exposure? To answer this question it was investigated whether the permeability of the cell membrane was altered in regard to biomolecules by vacuum treatment: after vacuum exposure the resuspended *E. coli* cells were centrifuged and the uv spectrum of the supernatant was measured (Fig. 3). The absorption in the region of 260 nm was increased compared with the controls. This means that the cellular membrane became permeable to substances absorbing at 260 nm, which is known to be a characteristic of nucleic acid compounds.

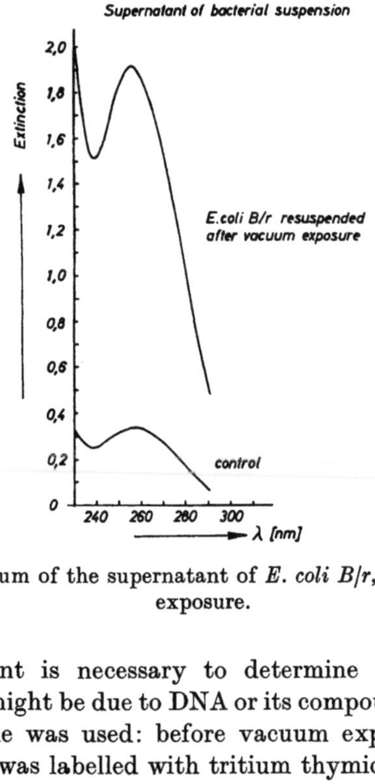

Fig. 3. Absorption spectrum of the supernatant of *E. coli B/r*, resuspended after vacuum exposure.

A second experiment is necessary to determine whether this increased absorption at 260 nm might be due to DNA or its compounds. Therefore a radioactive tracer technique was used: before vacuum exposure the DNA of the mutant *E. coli B/r* T^- was labelled with tritium thymidine (thymidine-methyl-^3H). After vacuum exposure the bacteria were resuspended and then filtered. The radioactivity of the filtrate was measured and compared with that of the controls; the measurement showed that the cell membrane became more permeable for the labelled DNA compounds after vacuum exposure, because the filtrate of the cells resuspended after vacuum treatment showed about three times more radioactive material than that of the controls.

From these two experiments it can be stated: the permeability of the cellular membrane to nucleic acid compounds such as thymidine is increased by vacuum treatment.

It may be of interest to compare these results with findings obtained on frozen bacteria: they also have an increased membrane permeability for substances absorbing at 260 nm [8], which was shown to be caused especially by RNA fragments [9].

In conclusion we may summarize the following effects on bacterial cells obtained by vacuum treatment:

Vacuum effects on bacteria

Loss of colony forming ability related to water desorption of the cells

Effects on colony formers	Effects on non colony formers
Prolonged lag-phase of subsequent growth	
Increased sensitivity to X-rays	
Increased sensitivity to uv rays	
no photoenzymatic repair	No cellular elongation
little or no excision repair	No phage production
Decreased sensitivity to heat	No respiration

Increased permeability of the cell membrane to nucleic acid compounds

These results are of interest in some aspects: they contribute to an understanding of the role of water in biological systems and processes, for instance in regard to the origin of life and life on other planets; furthermore the investigations contribute to the experiments of planetary quarantine.

References

[1] H. Bücker and G. Horneck, Life Sciences and Space Research VIII, 33 (1970).
[2] J. P. Brannen, Interim Rep., Sandia Laboratories, SC-RR-70-439 (August 1970).
[3] H. Bücker and G. Horneck, Biophysik **6**, 69 (1969).
[4] G. Horneck, H. Bücker and H. Wollenhaupt, Life Sciences and Space Research IX, 119 (1971).
[5] H. Bücker, G. Horneck and H. Wollenhaupt, Biophysik **7**, 217 (1971).
[6] G. Horneck and H. Bücker, Strahlentherapie **141**, 732 (1971).
[7] C. S. Cox and F. Baldwin, J. Gen. Microbiol. **44**, 15 (1966).
[8] M. L. Speck and R. A. Cowman, in: Freezing and Drying of Microorganisms, University of Tokyo Press, Tokyo 1969 (p. 39).
[9] T. Morichi, in: Freezing and Drying of Microorganisms, University of Tokyo Press, Tokyo 1969 (p. 53).

Life Sciences and Space Research X — Akademie-Verlag, Berlin 1972

HYBRIDIZATION OF PROTEINS UNDER ULTRAVIOLET LIGHT

M. A. Khenokh, V. P. Pershina and E. M. Lapinskaya
USSR

Solutions of human serum albumin (c = 1%), ribonuclease (c = 1%), and their mixtures were subjected to ultraviolet light. Filtration through Sephadex G-75 showed two additional fractions formed from serum albumin. One corresponds to photochemical cross-linking of this protein, while the other, with a smaller molecular weight than the initial, results from photodestruction. Concurrent with aggregation, photolysis destroys the tertiary (formation of SH groups, i.e. rupture of S—S bonds: quantum yield $\varphi_{SH} = 1 \times 10^{-3}$), and primary (accumulation of NH_2 groups and carbonyl compounds: $\varphi_{NH_2} = 0.44 \times 10^{-3}$) structure of this protein. The irradiation of RNA-ase solutions also leads to the formation of two additional fractions: one has a higher and the other a lower molecular weight than the original protein. The latter fraction shows no enzymic activity. The inactivation of the enzyme increases with the radiation dose. In all the fractions of the irradiated enzyme, the number of SH groups increases, indicating the breakdown of disulphide bonds ($\varphi_{SH} = 0.8 \times 10^{-3}$). The amount of amino nitrogen also increases, indicating the rupture of peptide bonds ($\varphi_{NH_2} = 0.14 \times 10^{-3}$). Gel-filtration of an irradiated albumin-RNA-ase mixture (1:1) yields a new substance with higher molecular weight, which exhibits enzymic activity toward RNA. The formation of this aggregate was confirmed by filtration through Sephadex G-150. The elution profile demonstrates that the resulting substance differs from the aforementioned aggregates of albumin and RNA-ase. It shows RNA-ase activity. It is suggested that the new substance is a hybrid between RNA-ase and albumin. The photo inactivation of the hybrid increases with simultaneous photolytic destruction of its tertiary (increase in the number of SH groups) and primary structures (increase in the amount of carbonyl compounds and amino nitrogen). These data bear on the chemical conversions of abiogenically synthesized protein-like substances in the primeval ocean.

A full paper will be published in due course.

Preparations for the Exploration of Mars

Life Sciences and Space Research X — Akademie-Verlag, Berlin 1972

BIOLOGICAL INSTRUMENTATION FOR THE VIKING 1975 MISSION TO MARS

H. P. KLEIN[a] and W. VISHNIAC[b]

[a] NASA, Ames Research Center, Moffett Field, Calif., USA
[b] Dept. of Biology, University of Rochester, Rochester, N. Y., USA

A brief introduction is given on why Mars is of interest from a biological point of view, along with an overview of the Viking 1975 mission. Details are given about the four biology instruments aboard the spacecraft and the experiments for which they are to be used. These are: the carbon assimilation experiment to determine whether the soil is biologically active, by incubation in presence of ^{14}C-labelled CO and CO_2 (known to be present in the Martian atmosphere); the label release experiment to detect metabolic activity by the release of radioactive CO_2 from ^{14}C-labelled simple organic substrates; the gas exchange experiment to detect biological activity by repeated gas chromatography analysis of soil samples; the light scattering experiment, where increase of scattering and decrease of light transmission would indicate the growth of organisms. Examples are given of data obtained with terrestrial soils in these experiments.

1. Introduction

In the late summer of 1975, two unmanned spacecraft, constituting the Viking mission, will be launched to Mars from earth. The first of these is now scheduled to leave in mid-August, and the second, about two weeks later. The spacecraft will travel approximately 460 million miles and arrive some 300 days after launch. The precise landing areas for the Viking spacecraft have not yet been selected, but one is tentatively destined to land in the northern hemisphere of Mars, and the other in the southern hemisphere.

Each spacecraft will consist of an orbiter and a lander which will be mated during the interplanetary trip and separate upon command from earth. After reaching Mars, the Viking spacecraft will first be placed into orbit in order to help in the selection of a final landing site for the lander. Aboard the orbiters will be television cameras for imaging, an infrared spectrometer for water vapor mapping, an infrared radiometer for thermal mapping, and radios and other associated instrumentation for data transmission to earth. Once a site has been selected for the lander, it will separate from the orbiter and enter the Martian atmosphere, making measurements of the atmospheric composition and structure with pressure and temperature sensors, accelerometer, a mass spectrometer, and a retarding potential analyzer.

The entire lander system will be heat sterilized prior to launch in order to prevent contamination of Mars by terrestrial micro-organisms. During the nominal 90-day lifetime of each lander, a number of scientific investigations will be conducted, in addition to the biological experiments. These include meteorology, molecular analysis, seismometry, imaging, and physical and magnetic properties. After landing, two cameras will scan the local area and transmit both close-up and panoramic images back to earth to aid in selection of sites for subsequent sampling of the Martian surface.

2. The Biological Experiments

Of the several experiments aboard the Viking lander, it is possible that biologically important information will be obtained from experiments not primarily designed for biological purposes. It may well be that the imaging experiment will either directly or indirectly indicate the presence of organisms in the vicinity of the Viking lander. In addition, analysis of the Martian atmosphere could supply highly relevant information. As has been pointed out by Lovelock [1], the coexistence of oxidized and reduced gases in the Martian environment (for example, if methane were found to be present in the presence of carbon dioxide on Mars) would strongly indicate the likelihood of a biological process. Such atmospheric analyses will be performed by the investigators involved in the molecular analysis experiments aboard the Viking mission [2]. This same group will be sampling the Martian surface for the presence of organic matter through the use of a combined gas chromatograph–mass spectrometer. The presence of simple organic compounds in Martian samples could be attributed to abiogenic processes, but it is clearly possible that the "mix" of organics in the Martian surface would be best interpreted on the basis of a Martian biota. The gas chromatograph–mass spectrometer thus represents a powerful tool in the Viking armamentarium in regard to the question of life on that planet.

The biological experiments aboard the Viking lander are to be conducted by a team of biologists*, who have proposed a biological package with four different experiments, based on different assumptions about the probable nature of Martian organisms. The advisability of performing several different biological tests on the same soil sample has been repeatedly pointed out [3, 4]. The biological package on Viking will consist of the "gas exchange" experiment, the "carbon assimilation" experiment, the "label release" experiment, and the "light scattering" experiment. For the biological experiments, soil samples will be acquired by the sampler aboard the Viking lander and will be distributed both to the biology experiments, and to the gas chromatograph–mass spectrometer equipment, by a common assembly system. The soil samples, sized to contain particles under 2 mm in diameter, will enter a hopper where they will fall by gravity into metering cups associated with each biological experimental test station. The rotation of the soil distribution assembly causes

* H. P. Klein, (Ames) team leader; W. Vishniac, (Univ. of Rochester) asst. team leader, light scattering; N. H. Horowitz, (JPL, Cal. Tech.), pyrolytic release; G. V. Levin, (Biospherics Inc.), labelled release; V. I. Oyama, (Ames), gas exchange; A. Rich, (MIT); J. Lederberg, (Stanford).

rotation of the individual test chamber assemblies by way of an engageable Geneva mechanism, dumping the soil from the metering cavity either into an active test cell, or into a control test cell, or into dump cavities (Fig. 1).

Each of the biological incubation chambers will be capable of being utilized several times during the 90-day lifetime of the lander, and it is planned to acquire three or four different soil samples for each experiment aboard each Viking lander. Incubation is to be carried out at 10° ± 5 °C. In the event

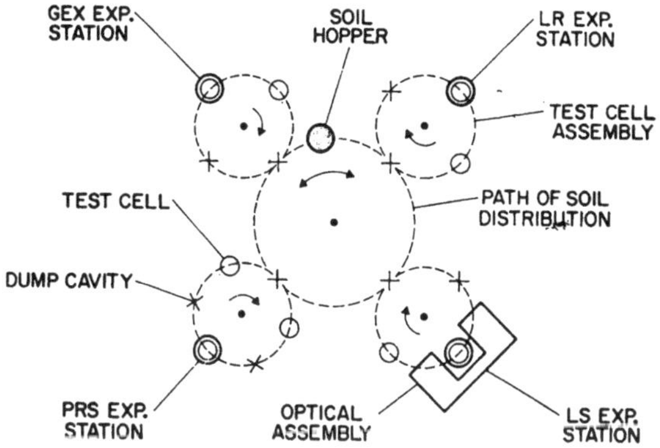

Fig. 1. Soil distribution system for biology experiments.

that any of the tests gives a positive response, controls will be performed on soil samples reserved for that purpose. For controls, it is planned to heat the soil samples to 160 °C for three hours before subjecting them to incubation and analysis.

The "carbon assimilation" experiment of Horowitz and coworkers [5] (also called the pyrolytic release experiment) takes advantage of the fact that the Martian atmosphere is known to contain carbon dioxide and carbon monoxide. In this experiment (Fig. 2), a soil sample is introduced into the incubation

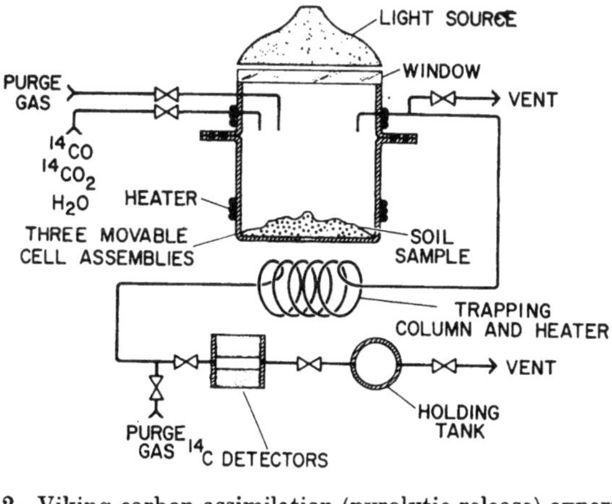

Fig. 2. Viking carbon assimilation (pyrolytic release) experiment.

chamber, purged with Martian gas, and then incubated in the light under ambient Martian atmosphere in the presence of small amounts of ^{14}C-labelled carbon monoxide and carbon dioxide. Assimilation of either of these simple compounds into organic matter will indicate the presence of biologically active material in the samples. The experiment is designed so that water vapor can be added upon command. A xenon lamp will supply the illumination, which will be filtered to remove radiation below 310 nm, in view of the findings of Hubbard et al. [6] that simple organic compounds are synthesized in the presence of low wavelength ultraviolet light from carbon monoxide. After incubation, the chamber is again flushed with Martian gas to remove unadsorbed CO and CO_2, and then heated to 600 °C to release any adsorbed radioactive gases. Under these conditions, organic matter is pyrolyzed to smaller fragments and, since the entire gas mixture is passed through a column containing firebrick coated with copper oxide, carbon dioxide and carbon monoxide will pass through, but organic fragments will be retained on the column. After the residual carbon dioxide and carbon monoxide have passed through the column, and the radioactivity falls to background levels, the column temperature is raised to 700 °C which now releases the bound organic material and simultaneously oxidizes these compounds and fragments to carbon dioxide. Thus, if either carbon monoxide or carbon dioxide had been assimilated into organic matter, this procedure would produce two radioactive peaks, one released at 600° and the other by heating the column to 700°. In the event that no carbon is assimilated, only a single peak would be observed. As is seen in Table 1, the results obtained with a breadboard model of this instrument demonstrate the two radioactive peaks that are obtained when an unsterilized soil sample is incubated in the light. Smaller peaks are observed when the same soil is incubated in the dark. The sterilized sample, however, gives only a single peak.

Table 1

Carbon assimilation experiment: test results

Procedure	Earth atmospheric conditions: 0.13 cm³ soil, 0.07 cm³ water, 63 hours incubation		
	Active Light	Active No light	Sterile Light
700° OVT elution after incubation atmosphere removed	190 DPM	321 DPM	280 DPM
600° pyrolysis (1st peak)	10^6	10^5	4×10^5
Post pyrolysis purge (OVT at 120°)	80	200	380
700° OVT elution (2nd peak)	1.3×10^4	3×10^3	360

OVT = temperature at which organic vapor trap is kept.

The "label release" experiment is an attempt to detect metabolic activity by the release of radioactive carbon dioxide from ^{14}C-labelled simple organic substrates. In this experiment of Gilbert Levin [7] soil samples are to be introduced into an incubation chamber, flushed with Martian gas and then incubated in the presence of ambient Martian atmosphere. At the beginning of incubation, a small volume of an aqueous mixture of radioactive substrates

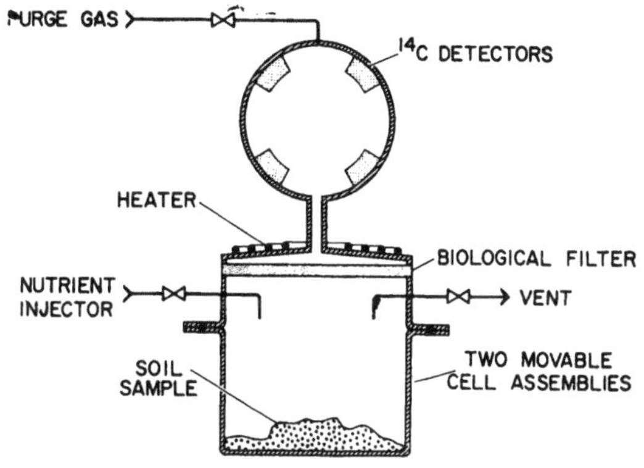

Fig. 3. Viking label release experiment.

will be added, and the headspace will be monitored for radioactivity (Fig. 3). The release of any radioactive gas (presumably CO_2) will be monitored by a suitable ^{14}C detector. At present, it is planned to incubate a 0.5 cm^3 sample of soil with 0.1 ml of nutrient medium. The composition of the nutrient mixture is not firmly settled. Much of the exploratory studies with a number of soils has been performed using a mixture of ^{14}C-labelled formate, lactate, glycine, and glucose. In such tests, radioactive gas is rapidly released while sterilized soils give a negligible response. In Fig. 4, results are shown, using a soil sample obtained from Death Valley, California. It is seen that the sterilized sample released some radioactivity, but the native material was orders of magnitude

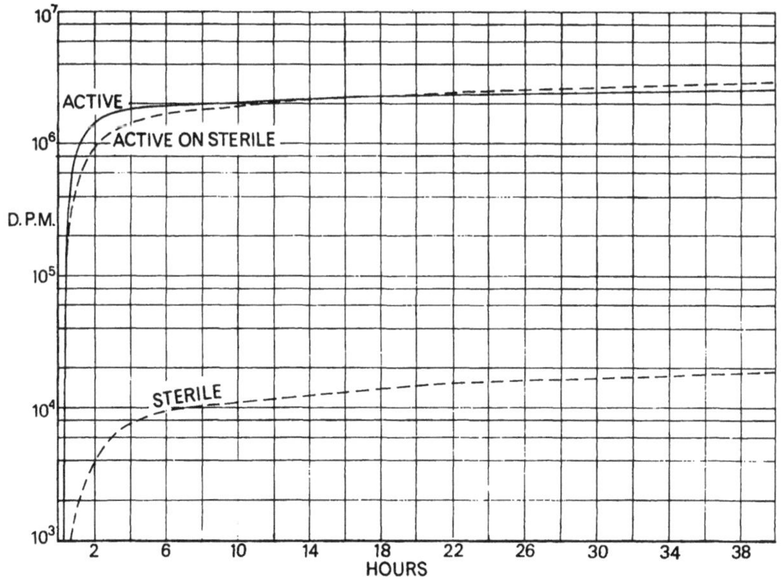

Fig. 4. Label release experiment; test results.

more active. In addition, an active soil placed upon a sterilized sample gave comparable results, demonstrating the "reusability" concept in these experiments.

In the "gas exchange" experiment of Vance Oyama [8], a 1.0 cm^3 sample of Martian soil is incubated with an aqueous medium containing a rich solution of organic material in the presence of the ambient Martian atmosphere (Fig. 5). Periodically, 100 μl samples of the enclosed atmosphere are analyzed by gas

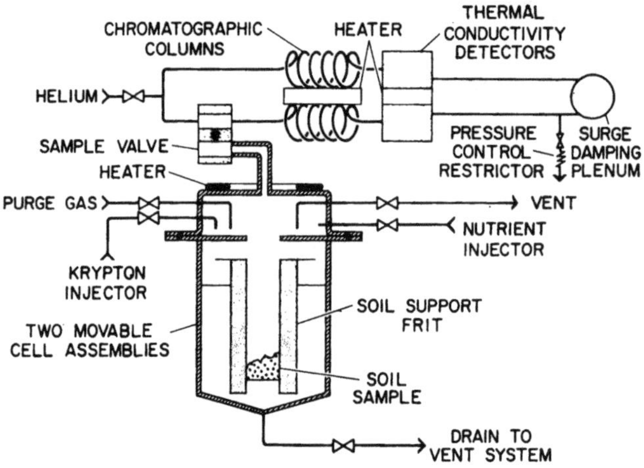

Fig. 5. Viking gas exchange experiment

chromatography for several gases over a 15-day period of incubation. Hydrogen, nitrogen, oxygen, methane, and carbon dioxide are easily resolved by the chromatograph system employed in this experiment, and either the uptake or release of gases can be monitored over the incubation period. The experiment

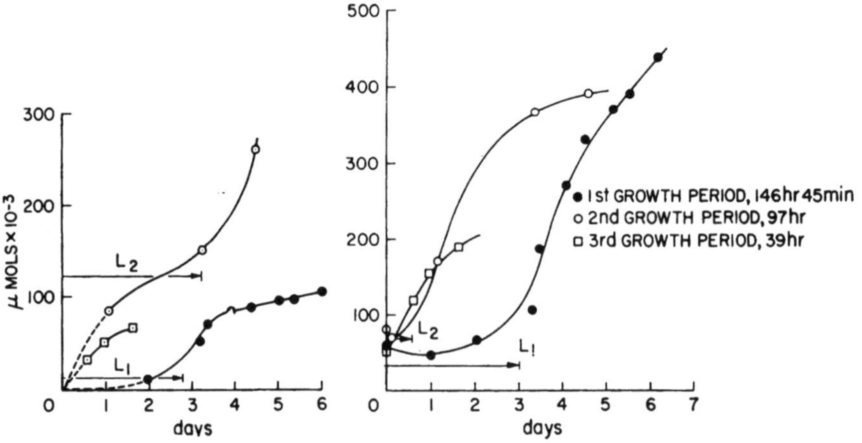

Fig. 6. Gas exchange experiment: test results; left, release of hydrogen on successive rechargings; right, carbon dioxide release.

is designed in such a way that the incubation compartment will be flushed, and the liquid drained after a 15-day incubation cycle, and then the original soil sample will receive a fresh charge of nutrients and ambient atmosphere. Gas analyses will be repeated a second time over a second 15-day period. Similarly, a third re-charging of nutrient and atmosphere is to be carried out. In Fig. 6, the release of CO_2 and H_2 in repetitive incubations of a sample of Bower's clay can be observed. In this case, as in most soils analyzed, gas release (or uptake) is slowest during the first incubation period, and then increase in subsequent re-chargings. In repeated tests, this procedure differentiates between soils with biological activity and those that have been sterilized. Occasionally, gas release has been observed in soils containing few or no organ-

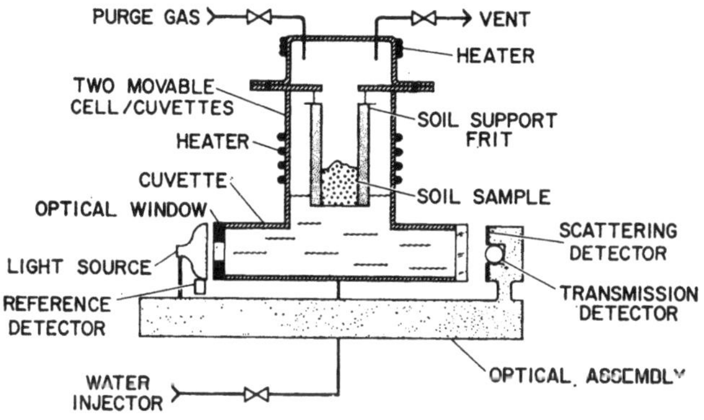

Fig. 7. Viking light scattering experiment.

isms, but in these cases, this non-biological activity can be differentiated from biological activity by the procedure of repetitive charges of fresh medium and fresh atmosphere. Using such repetitive incubations, artifactual release of gas in the absence of organisms decreases successively, and dramatically, after each fresh change of nutrients and atmosphere.

The final procedure, that of Wolf Vishniac [9], is one in which a portion of the Martian soil is to be incubated in the presence of distilled water and ambient Martian atmosphere. As shown in Fig. 7, soil is placed in the incubation chamber, which is separated from the liquid by means of porous filters (frits) designed to allow free diffusion of water, and movement of micro-organisms, but at the same time to retain the major portion of the soil sample. In preliminary tests, porous filters of approximately 100 μm effective pore size have proved to be most suitable for this role. Approximately 1.5 grams of soil will be used in the incubation chamber and about three to six times this volume of distilled water will be introduced from the bottom of the chamber to wet the soil. Indigenous extractable substances are expected to provide nutrients necessary for growth. The incubation chamber is designed so that a collimated beam of light will pass through the lower portion of the sample chamber. The signal will then be received by an assembly which consists of a small central detector on which the light beam is focused in order to obtain transmission measurements, and an outer detector area, which will receive a signal only when the light entering the optical path is scattered by particles. The combination of transmission and scattering measurements is expected to cover the range of particle densities from approximately 10^3 to 10^9 particles of 1 μm diameter per milliliter. Fig. 8 shows data obtained using an early breadboard model of this instrument. In this experiment, a sample of Holtville soil from

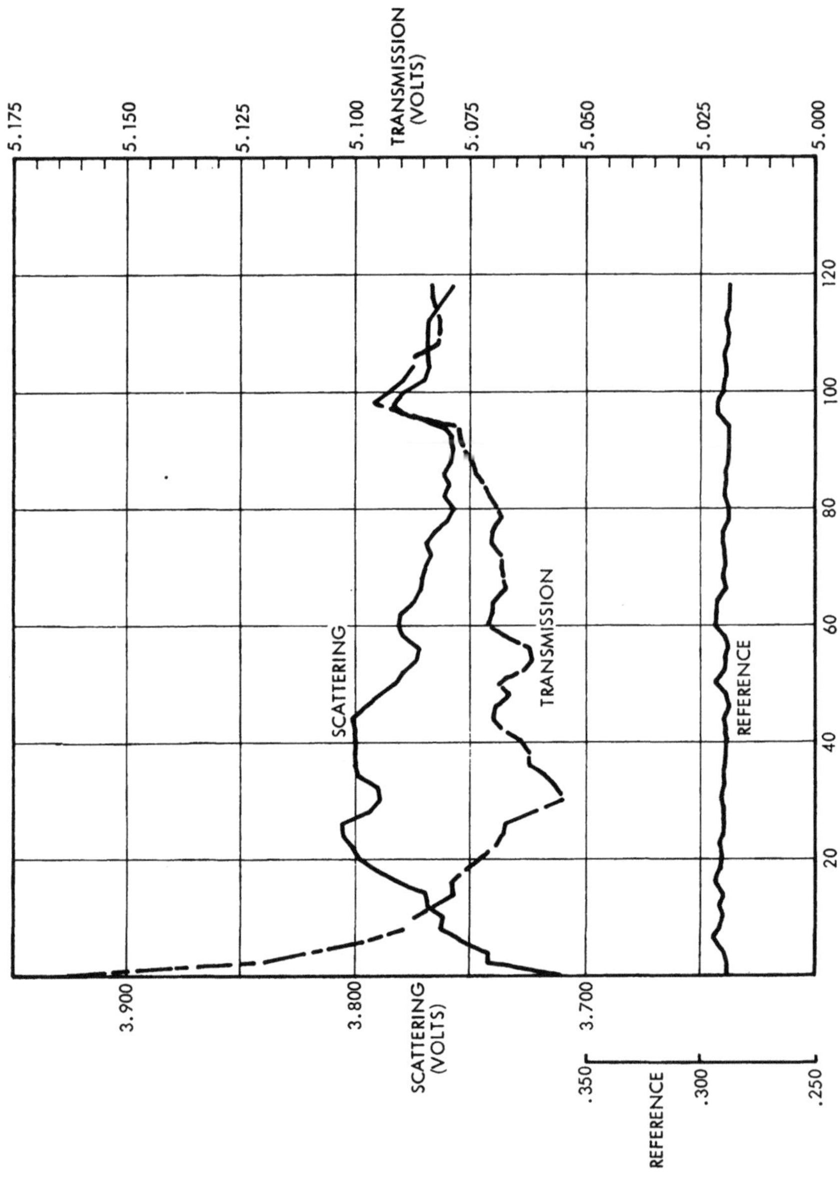

Fig. 8. Light scattering experiment: test results.

California was incubated for several days. During this period scattering increased and light transmission decreased, indicating the growth of organisms.

The combination of the four active biology experiments aboard each Viking lander covers a wide range of assumptions about the possible Martian biota. These range in gradation from dry to wet experiments, and from autotrophic to heterotrophic conditions. The combined integrated instrument, containing two beta particle detection chambers for counting ^{14}C-labelled materials, a thermal conductivity detector for the gas exchange experiment, and the optical system for the light exchange experiment, as well as nutrient injectors, gas storage facilities, heaters (for incubation as well as for sterilization) and the associated electronics for sequencing the entire operation, is being designed to occupy a volume of 1100 cubic inches, having the overall dimensions of 9.5 in. $\times$10 in. $\times$11.5 in. and to weigh approximately 20 pounds.

3. Discussion

From the biological point of view, the Viking 1975 mission can be regarded primarily as a test of the Oparin–Haldane hypothesis of chemical evolution. Findings of amino acids and hydrocarbons in the Murchison and Murray meteorites [10, 11], presumably originating in the asteroid belt between Mars and Jupiter, as well as the recent radioastronomical findings of formaldehyde, hydrogen cyanide, methanol, cyanoacetylene and other organic compounds in the interstellar medium [12, 13], all suggest very active chemical processes leading to the formation of compounds of great biological and biochemical interest. The molecular analysis experiments aboard the Viking 1975 spacecraft will be concerned primarily with a search for compounds of this type. On the other hand, the biological experiments ask the question whether such evolutionary processes reached a level of complexity characteristic of replicating systems. Biologists are exceedingly interested in this question and, consequently, in this opportunity to conduct experiments on the surface of Mars to look for evidences of living organisms. If evidence is obtained suggesting a Martian biota, this will clearly lead to new insights concerning current theories on the origin of life and on the fundamental properties of living systems. If, on the other hand, the data indicate a planet devoid of organic matter or of living organisms, this would lead to new speculations and theories about the origin and development of our solar system.

4. References

[1] J. E. Lovelock, Nature, Lond. **207**, 568 (1965).
[2] D. M. Anderson et al., Icarus, in press (1972).
[3] O. E. Reynolds and H. P. Klein, in: Proc. 2nd Int. Conf. on Environmental Problems of Man in Space, Springer-Verlag, Vienna and New York 1967 (p. 494).
[4] R. S. Young, R. B. Painter and R. D. Johnson, NASA Spec. Publ. No. 75 (1965).
[5] J. S. Hubbard et al., Appl. Microbiol. **19**, 32 (1970).

[6] J. S. Hubbard, J. P. Hardy and N. H. Horowitz, Proc. Nat. Acad, Sci. USA **68**, 574 (1971).
[7] G. V. Levin et al., Life Sciences and Space Research II, 124 (1964).
[8] E. L. Merek and V. I. Oyama, in: Life Sciences and Space Research VIII, 108 (1970).
[9] W. Vishniac, Aerospace Med. **31**, 678 (1960).
[10] K. Kvenvolden et al., Nature, Lond. **228**, 923 (1970).
[11] J. Lawless et al., Science, **173**, 626 (1971).
[12] L. E. Snyder and D. Buhl, Sky Telesc. **40**, 267 and 345 (1970).
[13] L. E. Snyder et al., Phys. Rev. Lett. **22**, 679 (1969).

Life Sciences and Space Research X — Akademie-Verlag, Berlin 1972

AN INTEGRATED MULTI-PURPOSE BIOLOGY INSTRUMENT UTILIZING A SINGLE DETECTOR, THE MASS SPECTROMETER

R. RADMER and B. KOK

Research Institute for Advanced Studies, Baltimore, Md, USA

A mass spectrometer is used to analyze the gas phase in a number of reaction vessels filled with Martian soil. By choosing appropriate incubation conditions this instrument can be used to perform a wide spectrum of experiments ranging from the observation of general indices of life, i.e. processes and patterns unexplainable by physico-chemical mechanisms, to assays utilizing isotopes which probe for specific metabolic processes. Of particular interest is the *in situ* incubation in which a Martian soil sample is maintained at a constant temperature and its gas phase composition analyzed with time. Properly interpreted, this is a very general life-detection probe which makes minimal assumptions as to the nature of Martian biology. Other assays and measurements concerning the soil and the atmosphere compatible with this method are also described.

1. Rationale

The paramount scientific problem involved in remote life detection is that there is no real understanding of what constitutes life. Consequently, it will be difficult to recognize Martian life, stimulate its metabolism, or predict which of many possible questions will result in measurable responses. Because of these uncertainties, it is probably best to attempt to "converge" on the (possible) Martian biological system by performing a graded series of tests ranging from very general "inferential" experiments to specific metabolic probes.

The choice and design of any single life detection assay involves a balance between universality and sensitivity. A general approach, such as the biologically oriented atmospheric analysis or *in situ* observations described here, would make few assumptions, would impose few perturbations on the system being studied, and could respond to a wide range of life activities, albeit with somewhat limited sensitivity. On the other hand, a specific approach, such as a search for a certain enzyme activity, will invoke specific assumptions and may impose severe environmental perturbations, although such an approach might be quite sensitive. Because the density of any life on Mars is likely to be low, it will be difficult to separate biogenic "signal" from non-biogenic "noise". Furthermore, since all life detection systems are basically quantitative, and therefore ambiguous rather than "yes or no", many data must be collected if a reasonable confidence level is to be attained. It will therefore be necessary to observe as many events and parameters as possible concerning Martian life and its environment; the cross correlation of these measurements will then yield a higher confidence level than that obtained from the sum of the individual measurements.

The measurement of metabolic activity (or the determination of its long-term effects) is probably the most promising attribute on which to base life detection experiments [1]. The most general experiment of this type is the determination of biologically induced disequilibrium of the atmosphere or of the soil–atmosphere interface (see § 4, last paragraph). It should be recognized that all planets, living or dead, will be in a state of thermochemical disequilibrium due to the presence of an energy source (the sun) and an energy sink (space). Because of this fact, attempts to distinguish between two states of disequilibrium, abiogenic or (at least partially) life induced, will always result in a certain degree of uncertainty. However, any extreme departure from chemical equilibrium would be good evidence for the presence of life. Life detection experiments which rely on the measurement of metabolic processes will yield data with distinct energetic and kinetic characteristics. (This is true both for experiments which rely on the detection of a single enzyme activity and those which measure the enhanced metabolic activity due to the stimulation of growth.) For example, the observation of a complex, energetically "uphill", reaction, such as photosynthetic CO_2 fixation, would be highly indicative of biological activity. However, since such a process requires some sort of energy coupling and, therefore, a good deal of cellular integrity, the probability of detecting this process is low. On the other hand, energetically "downhill" reactions (such as respiration) and isoenergetic reactions (such as isotope "scrambling") require only catalysis to occur at an appreciable rate. This catalysis could be provided by isolated enzymes so that the probability of observing these reactions is not so strongly dependent upon maintaining cellular integrity and/or the proper environment. However, these reactions are also apt to occur non-biologically, and consequently such a reaction is a somewhat ambiguous index of life.

The kinetic features of the observed changes in gas concentrations can also be used to interpret their biological significance. For example, the occurrence of autocatalytic and transient phenomena suggests the presence of biological activity, while the occurrence of fairly rapid first-order reactions in the initial phases of the incubation suggests reactions due to a readjustment of equilibrium (see below). More important, the cross-correlation of several simultaneous kinetic events can lead to conclusions which would be unwarranted if based on a single event or time course.

2. Technique

In addition to the scientific problems involved, there are considerable technical constraints which must be considered. The fact that these missions are limited in power, weight, etc., suggests that every instrument flown should be used for as many measurements as possible. Since the ability to manipulate the sample is similarly limited, it is probably easiest to measure gases. The mass spectrometer presently seems to be the instrument of choice for these experiments. It is simple to operate, flight type models are available, and it is capable of nondestructive kinetic analysis of many gases and their isotopes. These considerations suggest a life detection approach which utilizes a mass spectrometer to analyze, as a function of time, the composition of the gas phase in a number of ampoules filled with Martian soil. (Although all experiments described here were performed using a mass spectrometer with no ancillary analytical equipment, the overall system could be improved by the inclusion of either a gas chromatograph or selective gas filters.) In the proposed minimum configuration a mass spectrometer communicates with the atmosphere and a collection of ampoules ($\sim$1 ml each) which are filled with Martian soil samples and incubated as described below. The gas phase in these ampoules is repeatedly analyzed (with respect to an added inert internal standard such as Ar or He) during the period of incubation on Mars to detect "biologically induced" concentration changes. In the following paragraphs, we will briefly describe a graded series of measurements utilizing this configuration which is capable of obtaining a good deal of information concerning Mars, its environment, and its possible biology (for a more detailed description of these experiments see [2]).

3. The Detection of Life from Atmospheric Composition

Lovelock [3] has stressed the possibility of inferring the presence of life from the composition of the planetary atmosphere. For example, the coexistence of methane and oxygen in the earth's atmosphere is presumptive evidence for the presence of life. Although the non-catalytic oxidation of methane is slow, it is irreversible; there is no rational non-biological route for the production of methane in an oxidizing atmosphere. Similarly, no non-biological mechanism is known which might account for the present high abundance of O_2 ($\sim$20%) in the earth's atmosphere; the very observation of this high O_2 concentration suggests that a biological O_2-producing process exists (or existed in the past).

The gross composition of the Martian atmosphere has been studied by remote observation and found to consist largely of CO_2 and its dissociation products at a total pressure of about 10^{-2} atm [4]. However, it is still very important to determine the micro-constituents of the Martian atmosphere, a task which is probably beyond the capabilities of remote observation. If observers on Mars studied the earth remotely (as we have studied Mars) it is very unlikely that they would detect the small amount of CO_2 ($\sim$0.03%) in the terrestrial atmosphere. Finding no evidence for a CO_2 cycle, they might conclude that life did not exist on earth. Similarly, our present concept of the Martian atmosphere as a potential supporter of Martian life might change if we were to detect N_2. A detailed atmosphere analysis, in addition to being a general test for Martian life, would also help to define the Martian habitat as a possible environment and source of raw materials for biota, at least within our terrestrocentric biological experience. It might also help to set the stage for future experiments. In practice, a single mass spectrometer scan of the Martian atmosphere would not be sufficient to determine uniquely its constituents, particularly in the region around mass 28 (CO, N_2, and a fragment of CO_2). However, this difficulty can be circumvented by the use of a primitive gas chromatograph or filters designed to remove the interfering gases.

4. *In situ* experiments

In situ experiments (as we will use the term) attempt to observe and measure existing phenomena with a minimum of perturbation. As a class these experiments differ from the biologically oriented atmospheric analysis described above in that *in situ* experiments attempt to detect biologically induced changes

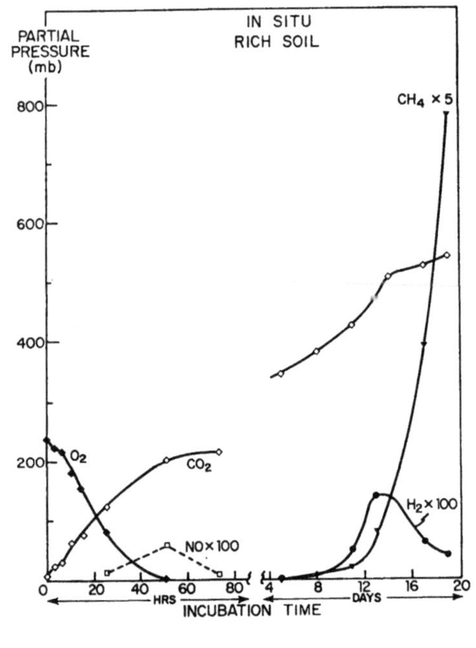

Fig. 1 A

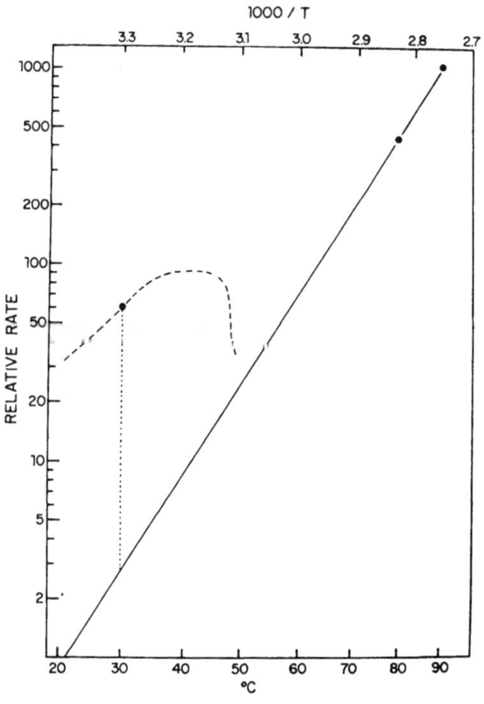

Fig. 1 B

in gas concentrations as a function of time. The static and kinetic analysis of the atmospheric composition above and below the soil surface is one type of *in situ* observation. For example, at a depth of 15 inches in dry forest ground we have observed that $[O_2]$ concentration was 3% lower than that of the atmosphere while $[CO_2]$ was ten-fold higher. In wet soil, $[O_2]$ was about 15% of the atmospheric concentration, $[CO_2]$ was five-fold higher, and methane was clearly detectable. Another *in situ* experiment involves the analysis of the gas phase enclosed in a transparent dome placed over the planetary surface. Since enclosure partially isolates the soil surface from the large atmospheric buffer, it is now possible to observe the diurnal succession of exergonic and endergonic metabolic events.

Sensitivity can be further increased at the price of deviating somewhat from the more ideal *in situ* experiments by transferring the Martian soil (plus its atmosphere) to a sealed vessel on the Mars lander. In this case one observes the sequence of metabolic events as the system runs energetically "downhill". An example of this last *in situ* experiment is illustrated in Fig. 1. In this experiment the gas phase changes were measured over an enclosed sample of forest soil. As oxygen disappeared CO_2 was evolved, and as the system became anaerobic, the transient production of NO (or N_2O) was observed. Upon further incubation hydrogen and methane were evolved. Fig. 1B illustrates the drastic changes in gas composition during a 50-day *in situ* incubation (note the appearance of peaks due to the formation of organic volatiles). Although these data are too complex to be analyzed into a set of unique constituents, the overall picture is a striking example of the anaerobic synthesis of many

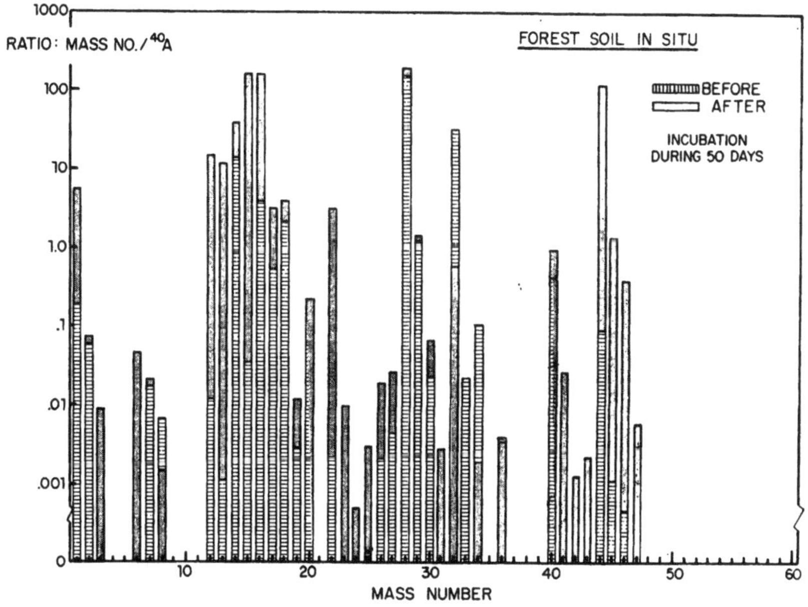

Fig. 1 C

Fig. 1. Changes in the gas phase above an *in situ* incubation of forest soil at 30°. A total volume of 250 ml enclosed 220 g soil having a plate count of 3×10^7 organisms/g soil and a water content of 16% wt/wt. In this and all succeeding experiments each observation is a mass ratio measurement; the peak height of the monitored gas is compared with the height of the mass 40 peak of argon used as an internal standard. A, time course; B, semi-logarithmic plot comparing the composition of the gas phase at T_0 and after 50 days incubation; C, Arrhenius plot of data obtained by incubating soil at three different temperatures (see text for details). The dashed line shows the hypothetical rate of biological O_2 uptake at different temperatures (adapted from Basic Bacteriology by C. Lamanna and M. F. Mallette, Williams and Wilkins Co., Baltimore). The dotted line relates the biological rate to the extrapolated non-biological rate at the same temperature.

organic compounds. With poor soil, far less dramatic changes were observed. However, after sufficiently long incubations we have never failed to measure substantial changes in the composition of the gas phase. It should be emphasized that, in general, the low pressure of the Martian atmosphere will enhance the sensitivity of this experiment due to the smaller concentration of diluting molecules.

The design and interpretation of a suitable control for this experiment presents some difficulty. Attempts to "sterilize" a sample usually result in other perturbations which sufficiently alter the sample so that it no longer serves as an adequate control. This is particularly true of heat sterilization, in all likelihood the only method available aboard a Mars lander. Alternatively, it may be possible to dispense with control measurements altogether, relying instead on the cross-correlation of many observations to arrive at firm conclusions. This approach would gain simplicity at the price of added ambiguity. For example, on earth the uptake of small amounts of O_2 with or without the concomitant release of CO_2 cannot be ascribed *a priori* to biological metabolism. However, if this same incubation mixture were later to become anaerobic and release H_2 and CH_4, it could be reasonably concluded that a biological system was present in the ampoule.

It might be possible to circumvent the difficulties inherent in a "sterile control", by using a procedure in which control samples are incubated and monitored at temperatures beyond the range of (most) biological activity. For instance, in the experiment illustrated in Fig. 1C, three terrestrial soil samples were incubated *in situ*, one at 30° and the others at 80° and 90°. Although all samples consumed O_2 and evolved CO_2, they did so by different mechanisms; the gas exchange at 30° was mainly due to respiration, while that at 80° and 90° was due to the non-biological oxidation of soil organics. The Arrhenius plot presented in Fig. 1C shows that in this experiment the background non-biological oxidation rate at 30° is $\sim 5 \times 10^{-3}$ times the rate at 80°, and that 96% of the O_2 uptake at 30° was due to respiration.

It should be emphasized that these nonbiological reactions reveal that the soil–atmosphere system of the earth is far from equilibrium, and thus an experiment of this type on Mars would be highly instructive. In fact, a similar type of experiment using differential thermal analysis has been suggested as a test for biologically induced disequilibrium [3].

5. Experiments which attempt to stimulate Growth

The most effective way to promote growth and metabolic activity in terrestrial soils is by the addition of nutrient materials. This approach can also be used in a search for extraterrestrial life, although in this case the choice of substrate will be difficult and critical. If this growth and/or metabolism is somehow reflected in gas exchange reactions, these processes can be detected using a mass spectrometer. No prior assumptions as to the identity of these gases are required.

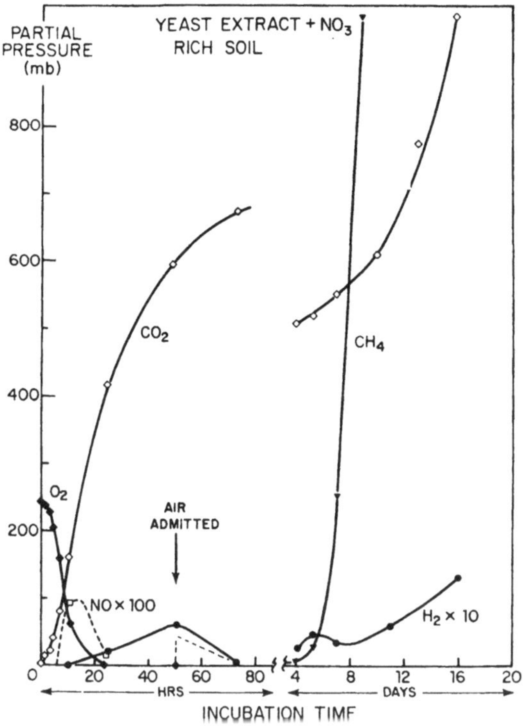

Fig. 2 A

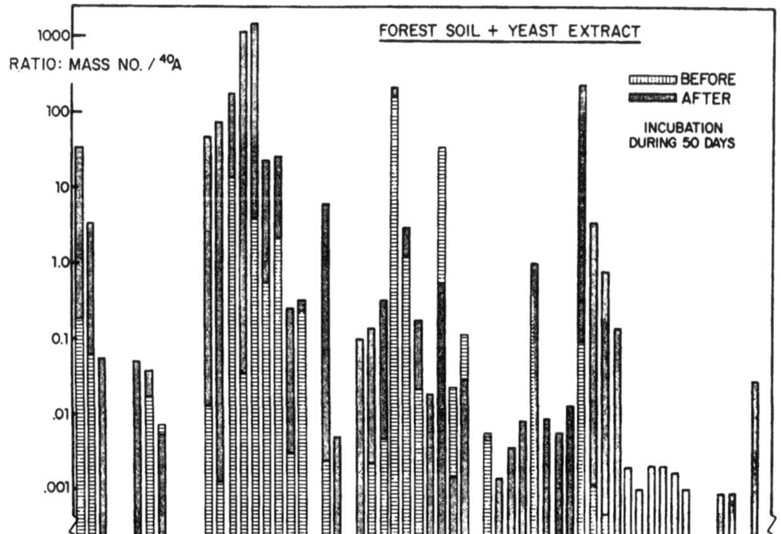

Fig. 2 B

Fig. 2. Changes in the gas phase above a sample of forest soil (see Fig. 1) incubated in the presence of yeast extract and KNO_3. 300 g soil plus 60 ml of a solution of 1% yeast extract – 10 mM KNO_3 was incubated in a 250 ml glass vessel. A, time course; B, comparison of the composition of the gas phase at T_0 and after 50 days incubation.

5.1. Addition of Water

Since water is presumably a rate-limiting commodity on Mars, its addition might enhance biological activity, provided of course that the added H_2O did not destroy Martian biota adapted to the very arid conditions. Alternatively, the H_2O could be added in the form of water vapor, in which case it might be less available to Martian biota but would not cause cell lysis. Experiments in which water was added to dry soil showed a large increase in biological activity, as measured by changes in the gas phase. The optimum H_2O content for aerobic metabolism was found to be about two-thirds of soil saturation, in accord with common soil microbiological experience [5].

5.2. Addition of Organic Substrate

Fig. 2 illustrates the results of an experiment in which forest soil was incubated with an aqueous solution of yeast extract. The gas compositional changes in relatively short periods of time are indicative of the intense and varied biological activity induced in the soil. A similar experiment using poor soil gave comparable results, suggesting that the biological sensitivity of this experiment is only a moderate function of the density of organisms originally present in the soil. In the event that true, earth-type "growth" does not occur, Martian biota (subcellular particles or enzymes) might still exhibit enzymatic activities and carry out partial conversions of the substrates and thus yield detectable signals of metabolism.

6. Specific Metabolic Probes using Stable Isotopes

In accord with the graded approach described in § 1 we have designed several metabolic probes which are compatible with the proposed experimental configuration. Although such assays have a low probability of success due to their specificity, the fact that experimental conditions are manipulated to generate and detect a specific spectrum of products can make a specific probe very sensitive for a given metabolic reaction. The following paragraphs describe a few of these probes.

6.1. $^{13}CO_2$ Formation from Organic Compounds

One of the most important terrestrial biological processes is the oxidation of organic compounds to CO_2 with a concomitant release of energy. This reaction is the basis of a life-detection method proposed for the 1975 Viking in which the release of $^{14}CO_2$ from terrestrial organic substrates is monitored by a radioisotope detector. With the replacement of ^{14}C by ^{13}C labels, this experiment can be performed using a mass spectrometer in place of a radioisotope detector. Fig. 3 presents data obtained in an experiment in which a rich forest soil was incubated with ^{13}C acetate. The extreme initial slope of the $^{13}CO_2/^{12}CO_2$ ratio illustrates the inherent sensitivity of this ratio measurement. An identical experiment using a very poor soil (about 10^4 bacteria/gram) showed a large signal within an hour. Control experiments in which the labeled acetate was

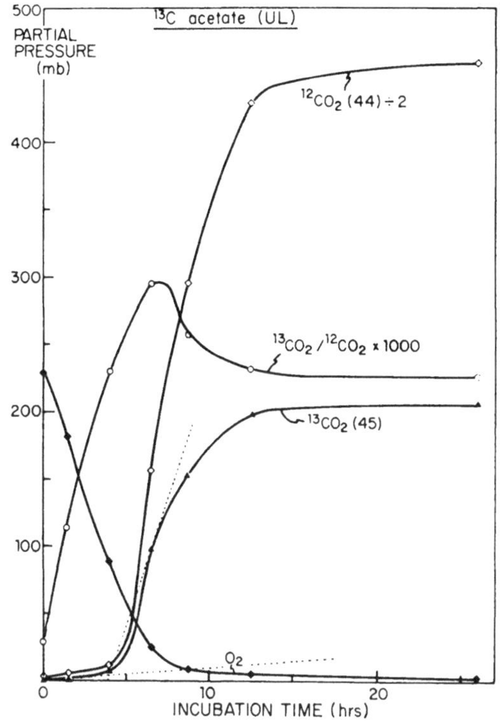

Fig. 3. The evolution of CO_2 by a sample of forest soil (see Fig. 1) incubated with ^{13}C sodium acetate. 100 g soil plus 20 ml of a solution of 10 mM sodium acetate uniformly labeled with ^{13}C (61.2 atom%) was incubated in a total volume of 60 ml. The dotted lines illustrate the initial and log phase slopes.

added either to heat sterilized soil or to water showed no increase of this ratio, suggesting that the changes shown in Fig. 3 were indeed life-induced.

6.2. Reactions involving Molecular Hydrogen

Many terrestrial organisms from widely different genera metabolize molecular hydrogen, either evolving it from reduced compounds and/or utilizing it for the reduction of oxidized compounds. In addition, many of these organisms and their cell free extracts catalyze the exchange reactions $D_2 + H_2 \rightarrow DH$ and $D_2 + H_2O \rightarrow DH + DHO$, often at a greater rate than the net throughput. This exchange or "scrambling" reaction is isoenergetic and does not require growth or cellular integrity.

Fig. 4 presents data obtained in an experiment in which soil was incubated in an atmosphere containing H_2 and D_2. Both net hydrogen uptake (as measured by the decrease in D_2) and the formation of HD were detectable after an incubation time of a few hours. Both of these rates — the rate of scrambling and the rate of net uptake — depend upon O_2 concentration. Although non-biological catalysts such as platinum and some copper compounds are able to activate hydrogen, such non-biological interferences can be corrected for by the use of a sterile control.

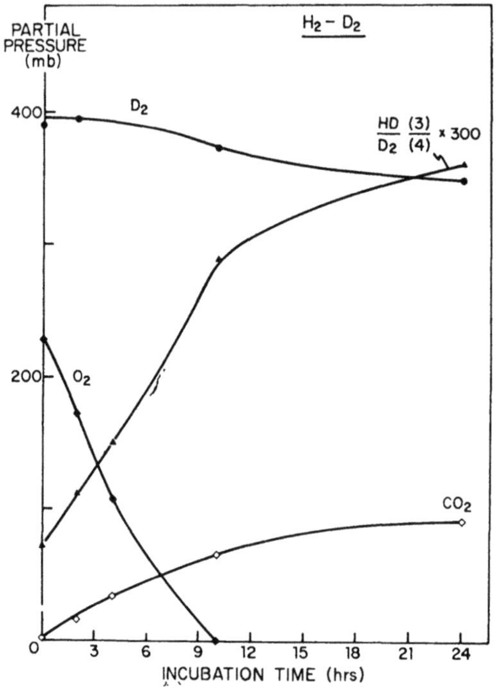

Fig. 4. The time course of hydrogen uptake and H_2 D_2 exchange in a sample of forest soil incubated in an atmosphere enriched in H_2 and D_2. 100 g soil plus 20 ml of a solution of 10 mM sodium acetate was incubated in a total volume of 60 ml. The initial gas phase above the soil in this experiment was approximately as follows: D_2, 40%; H_2, 30%; O_2, 20%; Ar, 10%.

6.3. Terminal Electron Acceptors other than O_2

Under anaerobic conditions, many terrestrial organisms are able to utilize oxyanions such as NO_3^-, SO_4^{2-}, and PO_4^{3-} instead of O_2 as terminal electron acceptors. On Mars, where the O_2 availability is very limited, this metabolic pattern might be prevalent. Two different approaches are available to detect the reduction of these oxyanions mass spectrometrically. In the first, the oxyanion is labeled with ^{18}O. Upon reduction, the ^{18}O will appear in water and can be monitored by the increase of ^{18}O in CO_2 with which it spontaneously exchanges its oxygen. In the second approach, the central atom of the oxyanion is labeled and the reaction monitored by the appearance of gaseous products, such as $^{15}N_2$, ^{15}NO and $^{15}N_2O$ from ^{15}N-labeled nitrate. Fig. 5 illustrates the results from an experiment in which $K^{15}NO_3$ was added to a soil sample. These data show that, especially in rich soils, denitrification is a very intense process. Note that the reduction of ^{15}N-labeled nitrate yields changes at several mass numbers due to isotopic forms of NO, N_2O, etc. Similar reactions also occur using other oxyanions as electron acceptors. In the case of SO_4^{2-}, the gaseous product H_2S could be measured at mass 34; in the case of PO_4^{3-} the gaseous

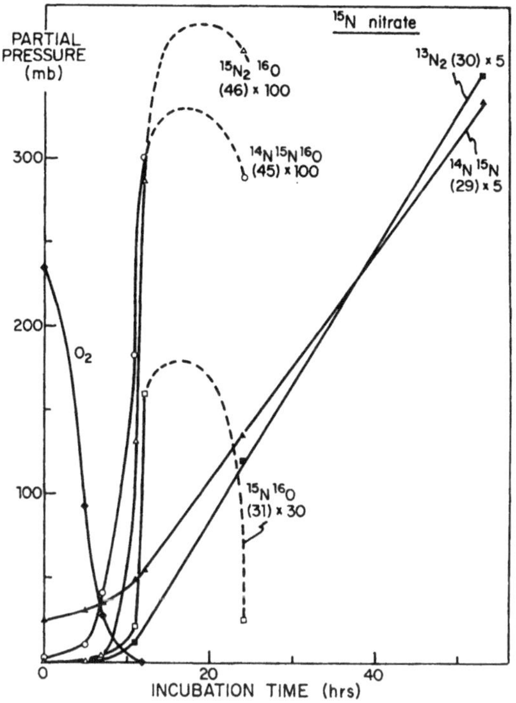

Fig. 5. The evolution of various labeled nitrogen compounds from a sample of forest soil incubated with $^{15}NO_3^-$. 100 g soil plus 20 ml of a solution of 10 mM sodium acetate–10 mM $K^{15}NO_3$ (99 atom% ^{15}N) was incubated in a total volume of 60 ml.

product, phosphine, PH_3, would also appear at mass 34. To date, we have not detected either of these products from soil samples. However, no concerted effort has been made to monitor either of these processes.

7. Some Factors governing the Ambiguity and Sensitivity of Biological Ampoule Experiments

In all experiments which attempt to detect "active biology", provision must be made to allow the discrimination of "true biology" from non-biological reactions which might mimic a biological system. In the following paragraphs we will describe several potential interfering reactions and their probable effect on the interpretation of the bioassays.

7.1. Photochemically induced Disequilibria

Hubbard et al. [6] have reported the photocatalytic production of organic compounds under conditions approaching those prevalent on Mars. Since O_2 is a possible constituent of the Martian atmosphere (as a photodissociation product) care must be taken to discriminate non-biological combustion from possible "Martian respiration". A parallel control procedure of the type described earlier (Fig. 1C) would allow this discrimination. However, the use of a heat-sterilized control would not provide an effective means for discriminating between these two processes. Oxygen would be consumed during the high-temperature sterilization and hence the incubation conditions of the viable and control samples would not be identical.

7.2. Gas Adsorption by Soil

Fanale et al. [7] have measured the degree of gas adsorption by pulverized basalt, a possible constituent of the Martian regolith, and found that this material has a very high internal surface area. Thus, it can be a good adsorbent. Although adsorption–desorption phenomena are essentially instantaneous [8] the accompanying diffusion processes are often slow enough to cause an appreciable time lag between any perturbation (such as change in gas pressure and/or temperature) and the resultant measurable change in gas concentration. Thus these adsorption phenomena may cause changes in gas phase composition on a time scale comparable with that of biological gas exchange. Consider the case of gases such as O_2, CO, Ar, or N_2, which have heats of adsorption $Q \sim 3.5$–4 kcal mole^{-1} [8]. If such a gas were in contact with 1 g of basalt with a surface area of 5.8 m^2 g^{-1} [7] at a pressure of 1 mmHg, about 3.6×10^{16} molecules would be in the gas phase and 3.1×10^{15} molecules, or about 10%, adsorbed on the basalt (see Appendix). Therefore if, for example, Ar were added to a system which had no competing species, roughly 10% of the gas would be adsorbed. It should be noted that this is a worst case; under conditions prevalent on Mars the Ar must compete for adsorption sites with H_2O and CO_2, both of which are much more strongly adsorbed than Ar.

7.3. Temperature-induced Disequilibria

Temperature-induced changes in the position of equilibrium must be considered when interpreting the results of ampoule experiments, and *in situ* experiments in particular. Significant temperature changes will result, for example, from the transport of the Martian soil sample to the thermostated lander and from the heating of samples for control procedures. The most reliable index to the possible effect of a reaction under a changing temperature is the value of its equilibrium constant. Reactions which have an equilibrium constant relatively close to unity usually show the greatest measurable effects of small temperature changes. As an example, consider the solubility of CO_2 in water. At room temperature, CO_2 is roughly equally distributed between the gaseous and aqueous phases, while at temperatures approaching the boiling point of water the CO_2 resides almost completely in the gas phase. Thus, assuming the gaseous and aqueous volumes to be equal, the resultant hundred-fold change in equilibrium constant results in two-fold change in CO_2 partial pressure. On the other hand, reactions which have extremely high or low equilibrium constants do not show an appreciable change in reactant concentration despite large changes in the calculated equilibrium constants with temperature. For example, although the combustion of formaldehyde (a possible product of a non-biological photochemical reaction [6]) shows a change in the calculated equilibrium constant from 10^{91} at 25° to 10^{75} at 90°, this change will not be reflected by any measurable changes in the composition of the gas phase.

Because adsorption–desorption reactions often have equilibrium constants relatively close to unity, a change in temperature can cause these phenomena to have a measurable effect on gas phase composition. Considering again a gas for which $Q \sim 4$ kcal mole^{-1}, calculations (see Appendix) show that roughly one half of the gas is desorbed when the temperature of the system is raised from 270 °K to 300 °K (−3 °C to 27 °C). The problem then reduces to one of kinetics, i.e. can these events occur on a time scale comparable with the "biological" processes of interest. Fanale et al. [7] observed that roughly two thirds of the H_2O (a strongly adsorbed species, $Q \sim 10$ kcal mole^{-1}) adsorbed on pulverized basalt was released within 90 minutes after desiccation at 29 °C. Since the time of molecular transport in an adsorption–diffusion process is strongly dependent on the value of Q, the desorption of most gases of interest (other than CO_2 for which $Q \sim 8$ kcal mole^{-1}) should be much faster. For example, a rough calculation for the desorption of Ar (or O_2, N_2 or CO) from clay yields a value of less than one second for the average time required by a molecule to travel 1 cm (see Appendix).

Several precautions can be taken to ensure that these adsorption–desorption phenomena do not unduly compromise the data. In particular, the fact that helium is not significantly adsorbed at the temperatures of interest ($Q \sim 100$ cal mole^{-1}) suggest that it would be the ideal internal standard, except that its use would eliminate the possibility of using D_2 (see H_2–D_2 experiment above). It should be noted that problems due to temperature-induced equilibrium shifts can be minimized by acquiring the soil samples at the warmest part of the Martian day. At this time, (i) the temperature of the soil will approximate to that of the lander, and (ii) all chemical reaction systems will be closest to equilibrium.

7.4. Factors affecting Biological Sensitivity

The sensitivity of the bioassays described above is a function of both biological and procedural parameters. For example, if the bioassay relies on growth and that growth has a generation time much shorter than the incubation time of the experiment, the detection limit can be one organism, whatever its size. In this case, the instrument performance is not critical. However, for methods which do not rely on growth, procedural sensitivities will be of crucial importance up to the point that non-biological interferences set the limit. The ultimate procedural limitations will depend on the type of measurement being performed. For bioassays which depend on the detection of newly formed compounds, sensitivity is limited by mass spectrometer cleanliness, sensitivity and dynamic range. On the other hand, those bioassays which rely on the measurement of concentration changes of gases present initially will be limited by the precision of the mass spectrometer ($\Delta S_{min}/S$) and stability (% per day). The requirement for high stability can be circumvented to a certain extent by the use of a calibrating gas mixture. The only other experimentally controlled parameter which affects the biological sensitivity is the volume ratio of soil to atmosphere in the sample ampoule. Consequently, biological sensitivity can be enhanced by completely filling the sample volume with soil, allowing a minimum headspace.

Probably the most important factor affecting the sensitivity of a particular measurement is the composition of the atmosphere in which the measurement is made. As an example, consider the measurement of respiration (O_2 + organic carbon $\rightarrow CO_2$). On earth, $[O_2] \sim 0.2$ atm and $[CO_2] \sim 3 \times 10^{-4}$ atm. Assuming a precision of ~1% and an adequate dynamic range, the minimum detectable change of $[CO_2]$ would be 3×10^{-6} atm and that of $[O_2]$ would be 2×10^{-3} atm. Ignoring solubility differences (which are quite significant), it appears that CO_2 evolution is about three orders of magnitude more sensitive an indicator of respiration than is O_2 consumption. However on Mars the opposite situation probably exists, i.e. $[CO_2] \sim 10^{-2}$ atm and $[O_2] \sim 10^{-5}$ atm. In this case O_2 consumption could be measured with 1000 times greater sensitivity than CO_2 evolution.

It is instructive to consider just how sensitive an experiment of this type can be under favorable circumstances. If we assume that 1 g of soil is incubated for 10 days under 1 ml of Martian atmosphere ($[CO_2] \sim 10^{-2}$ atm, $[O_2] \sim \sim 10^{-5}$ atm) and analyzed by a mass spectrometer having a precision of 1%, the minimum detectable (hypothetical) $[O_2]$ change would be 4×10^{-12} moles. This is the amount of O_2 consumed by 20 terrestrial bacteria in ten days (see Appendix). By way of comparison, surface samples of desert soils contain of the order of 10^5 organisms/g soil [9]. Although these calculated values cannot be taken too seriously, they do suggest that the ultimate detection limit for many of these bioassays will be set by non-biological interferences rather than by measuring limitations.

Appendix

Calculation of the Amount of Ar which can be adsorbed on Basalt (general reference: [8])

Consider a gas with a heat of adsorption $Q \sim 4$ kcal mole^{-1} (e.g. Ar, O_2, CO, N_2). The time of adsorption $\tau \sim 10^{-13}\, e^{Q/RT}$ where R is the gas constant and T the temperature in °K, whence $\tau \sim 1.6 \times 10^{-10}$ s at 0 °C.

The number of adsorbed molecules per unit area $\sigma = n \times \tau$ where n is the number of molecules striking a unit area of surface per unit time and is given by $n = 3.5 \times 10^{22}\, p/(MT)^{1/2}$ where p = pressure in mmHg and M = molecular weight. For argon at 1 mmHg and 0 °C $n \sim 3.4 \times 10^{20}$ s^{-1}, whence $\sigma = 5.4 \times \times 10^{10}$ molecules cm^{-10}.

The pulverized basalt studied by Fanale et al. [7] had a surface area of 5.8×10^4 cm^2 g^{-1}; assume the ampoule contains 1 ml of gas phase and 1 g pulverized basalt, then under the conditions described above the total adsorbed gas $\sim 3.1 \times 10^{15}$ molecules g^{-1} basalt. The gas phase, contains 1 ml of gas at a pressure of 1 mmHg, or $\sim 3.6 \times 10^{16}$ molecules; therefore under these conditions the adsorbed gas is roughly 10% of the total gas in the ampoule.

Calculation of the Effect of Temperature on Adsorption (general reference: [8])

Assume a gas with $Q \sim$ 4 kcal $mole^{-1}$ such as argon (see above) is adsorbed at 270 °K (−3 °C). At 270 °K: $\tau \sim 1.7 \times 10^{-10}$ s, and $n \sim 3.4 \times 10^{20}\ s^{-1}\ cm^{-1}$ so that $\sigma \sim 5.8 \times 10^{10}\ cm^{-1}$. If the temperature is now raised to 300 °K (27 °C), $\tau \sim 8.1 \times 10^{-11}$ s, $n \sim 3.2 \times 10^{20}\ s^{-1}\ cm^{-1}$ and $\sigma \sim 2.6 \times 10^{10}\ cm^{-1}$. Therefore a temperature excursion from 270° to 300 °C results in the desorption of about half of the adsorbed gas.

Calculation of the Time of Diffusion of Argon through Dry Clay (general reference: [8])

The average time for a molecule to pass through a capillary is $\bar{t} = l^2/2\, d\bar{u} + l^2\, \tau/2\, d^2$ where τ is the time of adsorption, l the length of the capillary, d the diameter of the capillary and $\bar{u}$ the mean velocity of the molecules given by $\bar{u}\,(8\, RT/\pi\, M)^{1/2}$. For argon at 300 °K, $\bar{u} \sim 4 \times 10^4\ cm\ s^{-1}$ and $\tau \sim 10^{-10}$ s (see above). If we assume the clay we are dealing with has a particle size $\sim$0.002 mm in diameter and a calculated pore diameter $\sim 10^{-4}$ cm, then $\bar{t}$, the calculated average time to travel 1 cm is $\bar{t} \sim 0.1$ sec.

Calculation of the Amount of O_2 consumed by One Bacterium

Assume that a bacterium consumes 200 µl (STP) O_2 (mg dry wt)$^{-1}$ hr^{-1} (This is a representative value, cf. *Handbook of Biological Data*, W. B. Saunders Co., Philadelphia and London).

200 µl (STP) $hr^{-1} \sim 2 \times 10^{-3}$ moles/10 days; 1 mg dry wt $\sim 10^{10}$ bacteria; therefore a bacterium consumes $\sim 2 \times 10^{-13}$ moles O_2 in 10 days.

References

[1] R. S. Young et al., NASA SP-75 (1965).
[2] R. Radmer et al., Science, **174**, 233 (1971).
[3] J. E. Lovelock, Nature, Lond. **207**, 568 (1965).
[4] C. A. Barth et al., Science **165**, 1004 (1969).
[5] D. Pramer and E. I. Schmidt, Experimental Soil Microbiology, Burgess Publishing Co., Minneapolis 1965.
[6] J. S. Hubbard et al., Proc. Nat. Acad. Sci. USA **68**, 574 (1971).
[7] F. P. Fanale et al., Nature, Lond. **230**, 502 (1971).
[8] J. H. de Boer, The Dynamical Character of Adsorption, Oxford University Press, London 1953.
[9] R. E. Cameron, in: Biology and the Exploration of Mars, Pub. 1296 NAS-WRC, 1966 (p. 164).

INDEX OF AUTHORS

Bei Fragen zur Produktsicherheit wenden Sie sich bitte an:
If you have any questions regarding product safety,
please contact:

Walter de Gruyter GmbH
Genthiner Straße 13
10785 Berlin
productsafety@degruyterbrill.com